COME TOGETHER: CREATION AND EVOLUTION JOINED

GLENN S. HAMILTON

Come Together: Creation and Evolution Joined

First Published September 2005 by Exposure Publishing
an imprint of Meadow Books:
35 Stonefield Way, Burgess Hill, West Sussex, RH15 8DW, UK

Printed simultaneously in the United States of America
and the United Kingdom.

All Scripture quotations taken from the
HOLY BIBLE KING JAMES VERSION.

British Library Cataloguing-In-Publication Data
A Record of This Publication is available from the British Library

Hamilton, Glenn S.
Come together: creation and evolution joined.
by Glenn S. Hamilton. – 1st ed

1. Science--Religion. 2. Creation--Evolution. 3. Social--Debate.
4. Biblical--Scientific.

ISBN: 190536329X

Table of Contents

Introduction

Since the mid-1800's, when Darwin's Theory of Evolution was unleashed upon the world, people have been debating the ideas of Creation and Evolution. The Creationists vs. Evolutionists phenomenon has been a frequent stumbling block and point of contention for educators, government, and the general public alike. State legislatures have debated pasting disclaimers inside the covers of textbooks, school boards are dictating what can and cannot be taught and how it is to be taught in our schools.

The often referred to Tennessee vs. Scopes, or 'Monkey Trial' of the twenties, was the initial test of laws designed to ensure that Creationism be taught in public schools. Now, only a little over eighty years later, we have evolved to laws prohibiting the teaching of Creationism. The debate is becoming even more intensive as I am writing this as the subject is all over the news. Through much research of both sides of this hot topic, the main points of contention have been examined carefully and an agreeable way to finally join the sides in this debate has been found.

Do not jump to any conclusions yet. I am first a Christian and second a scientist. I am not now, and never have been, a scientologist. I have earned the right to be referred to as a Christian by the process of being born again in Jesus Christ. I have earned the title of Scientist through many years of study and research and my attainment of the Bachelor of Science degree. Inspired by the struggles I faced for many years with the conflict found between science and the Bible, I have dedicated several years to building a convincing set of ideas that I am sharing through this book. Through many thousands of hours of research, prayer, and a little soul-searching I have arrived at some definitive conclusions that are supported by both Biblical and scientific principals.

This work is intended for those on both sides of the issue whether you are an atheist, theologian, Christian, scientist, or you just don't know what to think. It is especially for those of you that are in the position I once was: in the middle. How can a Christian be in the middle? Like I said, I am a scientist too. Having been raised a Christian and continuing that to the present and then spending years in the higher education system, I have been exposed to both sides of this issue.

There were times when I was confused and even repulsed by what was being laid down as fact within the classroom when it directly attempted to refute my faith. The most perplexing problem that faces all of us is how to seal the chasm that has grown between the sides. We don't want our children being taught that something is a fact when it goes directly against what we have been teaching them. The only solution is to carefully examine both sides in detail, attempt to sort out fact from fiction, and develop a theory that is sound both in Biblical and scientific terms. In order to accomplish this both sides need to take a deep breath, relax, and give me a little of your time. My most daunting task is to deliver this in as unbiased a manner as is possible. I believe that I have accomplished exactly that. So sit back, share a few hours of your time, and get ready for the Bible and science both to agree on Creation and Evolution.

About the Author

Glenn S. Hamilton describes himself as a Christian first and a scientist second. He has earned the right to both titles as a born again Christian, the Bachelor of Science degree he has earned, and his continuing studies and research as a graduate student. Glenn has done extensive coursework and research in the areas of Biology, Chemistry, Physics, Psychology, Education, and History. He is also widely read with the Bible being his 'primary life text'. Many other works of non-fiction on subjects such as science, education, history, and religion also rank among his favorites.

Glenn is currently pursuing a Masters degree at the University of Central Oklahoma in Edmond, Oklahoma. He has written articles for an encyclopedia to be published late in 2005. His current projects include various works on current controversial subjects and research in preparation for a comprehensive work about the history of science. His early education includes the public schools of Elyria, Ohio, where he was also born and raised, along with a year of study at the Elyria Christian Academy. Glenn, his wife Laura, and their 9-year-old son Joseph reside in Edmond, Oklahoma.

Chapter I

Definitions: Faith and the Nature of Science

Opening note: This work is intended for a vast range of readers who possess widely varying reading skill levels. The first chapter, while still essential for all readers, serves as a basic primer to not only introduce those with a minimal level of knowledge of scientific terms, but to also define the contextual definitions as they pertain to this book.

Central to an understanding of the issues on both sides of the Creation/Evolution debate are two critical areas that must be understood in their depth before any progress can be made toward a solution of any kind. Certain key terms must be defined and understood in order for anyone to be able to grasp the language necessary if an intelligent and informed observation is to be made. These two broad areas that we will start with will, for the purpose of this discussion, be named "Faith" and "The Nature of Science".

Background and Qualifications of the Author

First I feel that it is important to reveal my background and beliefs and reveal my qualifications that permit me to address these issues to allow you to study this issue with the same open-mindedness that I approached this project with. I am a 37-year-old graduate student with a Bachelor of Science degree from the University of Central Oklahoma. I have performed extensive study and research in the fields of Biology, Chemistry, Physics, Education, and History. Currently pursuing a Master of Adult Education degree, I have completed nearly half of the requirements.

My interest in the Creation vs. Evolution debate came about with my own struggle through the issues. I have always been first and foremost a Christian and have accepted Jesus the Christ as my Lord and savior. I am secondly a scientist having earned my Bachelor of Science degree and through the pursuit of continuing postgraduate studies and research. Do not jump to a hasty conclusion here: I am not, and have never been, a Scientologist.

I was raised in a Baptist church and now enjoy services nearly every Sunday at the multi-campus Life Church. I study my Bible often and firmly believe in the power of prayer. Having provided this background I feel that I have earned the right to share my discoveries with those on both sides of this issue. The hardest struggle that I have faced completing this work is that I want this to be readable by both sides of this issue.

It has been a huge struggle to create a work on such a debated topic and keep bias to a minimum. I, like most Christians, have a strong conviction in my faith. At the same time I am a researcher and a scientist. These facts do not permit me to simply discard scientific matters that have been tested and retested many times over using the scientific method. By the same token, I am very strong in my faith and will not blindly accept ideas that refute it.

I truly believe that I have not compromised my faith, in any way, by creating this work. I have consistently applied scientific standards and Biblical support tests throughout this work, and in particular, to my search for a new set of theories that can bring the sides together. My primary goal is to create a platform from which each side can respect the other, and eventually work together in ways that have never been conceived or thought possible.

The Sides and the Struggle

The Evolutionists frequently state that Creationists are a closed-minded group that is unwilling to even consider any other viewpoints. The Creationists feel that Evolutionists are anti-God in their beliefs. I believe that neither side is correct in these observations of the other. I first experienced my struggles with this issue when I began my academic career more than ten years ago. How could Creation be aligned with the scientifically based ideas of evolution or evolution be conceivable in the minds of the faithful? How can I maintain complete and total faith while trying to reconcile two irreconcilable ideas? While I could see no lack of people of faith in the academic institutions, I could see many who struggled with what they had to learn or teach. I have also witnessed many an aspiring doctor or scientist abandon their first choice of study and career for one that did not require years of learning theory that was unacceptable to their faith.

Let us look at some definitions essential for the purpose of our discussion.

What is Faith?

Faith is the belief in something without any solid evidence or proof that this something exists or is even real. There is proof that many people wrote the texts contained in scrolls that, when assembled, make up the Bible. The Bible has existed in what is essentially the same form for hundreds of years as the King James Version. Today there are many versions based upon the King James Version such as the Living Bible and the New King James Version, to name only a few of the more popular ones.

No matter which version of the Bible one happens to study or prefer, there is a recurrent theme: Faith is the vital ingredient for one's relationship with God. Faith is mentioned in both the Old and New Testaments and even by Jesus Christ Himself. Were the writings contained within the Bible Divinely inspired? In other words did God inspire these people who wrote the original manuscripts that make up the Bible to write what they did? True believers have to answer this question based solely upon faith. There exists no tangible proof that divine inspiration was involved at all in the creation of the original manuscripts.

On one side of this issue we have individuals that believe in something without any proof or evidence that it does indeed exist and on the other there are those who sometimes put too much faith in accepting other's views without any investigation or experimentation on their own. For an understanding of both sides by both sides everyone must understand the 'nature of science'

The Nature of Science

In order to clearly establish the definitions concerning the nature of science we shall use an illustration, or parable, utilizing something that everyone can see and agree upon: a lion cub. For the purposes of this part of the discussion it does not matter the sex of the lion – it is simply a lion cub. We can all agree that an individual that has not been long on this earth is considered a baby and for the sake of this discussion we will say that this lion baby, or cub, is one month old. At one month old let us assume that this lion cub has the same degree of motor development as a one-month-old human baby. While this assumption is probably not true, for the purposes of our discussion and the benefit of those that are not familiar with lion development, let us just use this assumption.

So here is our lion cub that cannot walk, lies around, purrs, and enjoys regular feedings. How do we know that this lion is alive? We can observe the lion and hear that it is making purring noises and see that it is moving its legs around. We can also see that this lion's eyes are open and looking around at the surroundings. We could go further and either feel for a pulse or maybe use a stethoscope to listen to this lion's heart and therefore identify beyond a shadow of a doubt that the lion is indeed alive.

What we have just done to this lion is called observation and experimentation. We were asked a particular question: "How do we know that this lion is alive?" We had to first look at the lion and see that it was moving. We listened and heard 'purring' noises that are characteristic of new born lion cubs. We even performed experiments and felt for a pulse and listened to the lion's heart with a stethoscope to determine that it was, in fact, alive. Could the lion make noises, move its limbs, or have a pulse if it was dead? "No" is the only obvious answer here.

What we have just done is to perform experiments and observe this lion to determine whether or not it is alive. We have collected evidence, or the results of our observations, that support our idea or hypothesis. The observation, experimentation, and the study of the results is the core of science itself. So let us now define observation as looking at, or studying something, to determine what its characteristics are. We can also define an experiment as a test like taking the lion's pulse or using a stethoscope to listen to its heartbeat.

Another term, evidence, is defined as the results of the observations and experiments that support or refute an idea. If the results support our idea then they are evidence that our idea is true. If the results do not support our idea then we have evidence that our idea is false. This does by no way mean that our assumptions were right or wrong; it only means that, so far, the observations and experiments performed either support or refute our ideas until further tests or observations are made.

It is the nature of science to be flexible not only with the results of further observations and experimentation but also with time and newer techniques that come along with progress. Phenomena that years ago were thought to have been explained satisfactorily have frequently been redefined later on by new and improved observational and experimental techniques. This process will continue, as time passes, due to progress in scientific methods.

Let us look at the example the process of disease. In ancient times people believed that illness and disease were brought on oneself at the whim of a particular god when an individual did not please them or angered them in some way. We then learned that certain contagious ailments could be passed from one person to another. Later the concept of germs came into our understanding which brought on the advent of various antiseptic techniques such as simple hand washing and sterilization of instruments to kill germs and prevent the spread of diseases. What was at one time thought to be true has been tested using new methods and has been found to be false. At the same time these new methods have brought us new possible answers and methods that keep ideas, hypotheses, and theories current and ever-changing along with our rate of progress.

Let us now continue on with our lion cub illustration to provide some additional insight into the nature of science. The first thing that we observed is that the lion was making purring noises. Could this be determined from across the room? Not with absolute certainty. You would have to be very close and observe that the lion's mouth was possibly open and maybe even see its chest or abdomen move slightly in sync with the noise to be absolutely certain that the lion did indeed produce the noise. Would this be evidence that this lion is alive? Most would think so but it is the nature of the scientific mind that it must probe further.

What about observing that the lion is moving its legs about – is this also evidence of a living lion? What I am trying to accomplish with this scenario is to provide a deeper understanding of the nature of science. A scientist must study the lion and gather numerous observations together – and perhaps even perform some tests - like feeling for a pulse – to determine the answer to the original question.

Flexibility of Science

Science is flexible and subject to change at any time and at any location. As we have been discussing, scientists observe, perform experiments, and collect information from those observations and experiments to support or reject an idea. These ideas are called hypotheses in the plural or hypothesis in the singular. A hypothesis is simply an idea or opinion that a particular observation will support or refute an idea. Let's go back to our lion for another illustration. The observed movement of the arms and legs, we have found, supports the idea that the lion is alive. To build upon that we have also added that the lion is making purring noises, moving its eyes about, and has a pulse.

Each added observation adds to the support of the idea that the lion is indeed alive and strengthens our hypothesis. One of the purposes of science is that it attempts to refute the correlation or relationship between these observations and ideas. Through constant testing of these observations and ideas science sometimes finds an alternate possibility that can explain a relationship.

In most cases the additional testing only strengthens the support of these relationships, especially in modern science with the tools and tests it possesses. There always has to remain the possibility that new methods and techniques can provide either a supporting result or a refuting one. Further experimentation and observation then follow to create either stronger support for our idea, a modification of it, or it can cause the discard of the idea and an entirely new one will take its place.

Let us look at an example of a change or modification that occurs frequently in science. If we continue along with the illustration of our lion we can say that time after time if a lion moves its legs it is alive. After being observed many times in different situations with different lion cubs and by different people and the results are always the same the hypothesis can become a theory.

Theories are ideas or a hypothesis that have withstood the tests of time and repeated testing over and over and the same results are always obtained. A theory, however, is always subject to change or modification as new supporting or refuting observations are made. We will progress along the line further and discuss laws for a moment but for the purposes of this discussion they will not be considered other than to define the term.

Laws are theories that have withstood many attempts to disprove them through experiments and observations and have never been refuted or changed. There are few of these in science but some examples are the law of gravity and the laws of thermodynamics. Laws, for the most part, define the behavior of objects from a physical perspective. All the physical laws that govern our universe are unique and complex in the way that they work together. These laws are so complex that they had to be designed by an intelligent Creator – God.

Faith in Science

Faith does have a place in science such as at the instance where an original idea is formulated and someone sets out to both support and refute it through observation and experiments. The scientist has to have a certain amount of faith to believe something that has never been tested before will turn out a certain way based upon their initial idea of the outcome. There is usually a hypothesis or preconceived idea about how things will turn out.

One way to reinforce the idea that science is always open to change of ideas and hypotheses is to look at the concept of the null hypothesis. A scientist will always formulate what is called a null hypothesis which says that there is no relationship between whatever is observed or experimented with. For example, I may be testing a new chemical that I hypothesize will kill a particular type of germ. My hypothesis is that if I apply the chemical to a sample of these germs they will die. I also have to formulate a null hypothesis that would state "The application of this chemical will have no effect upon the germ sample." The null hypothesis is necessary exactly because science is flexible and subject to change. The null hypothesis leaves open the possibility that the investigator's idea, or preconceived notion, may be wrong.

Let us go back to the example of our lion cub. At this point I have to stress again that an open mind must be present to consider our discussion as at this point we will get somewhat graphic with our hypothetical lion. This time we observe the lion and find that there is no movement of the limbs, the eyes are closed, we hear no noises, and check for a pulse and find none. At this point we even get our stethoscope and find that we can hear no heartbeat either. When we first observed the lion and noticed that the legs were not moving what are the possible explanations of this?

The first idea that would come to most is that the lion may be asleep but, at the same time, the idea that the lion is dead is valid also. Further observation is necessary to determine whether the lion is alive or dead and so we look further. In this absence of supporting observations that indicate that this lion is alive we must go further with our observations and maybe even conduct some tests or experiments to build a body of evidence that either supports or refutes the idea that the lion is alive or dead.

To go a little further with this lets say that we have determined that the lion is dead after all the observations and tests support the idea that this is so. We cover the lion with a sheet and leave the area. Let us assume that when we left the lion its legs were bent up against its body and it was lying on its side. We return the next day to find that the lion cub's legs are pointed straight out from its body! The lion must be alive!

We believe the above support since we discussed earlier that movement of the legs indicate that the lion is alive. We make further observations and see that the eyes are closed, there are no noises, and there is no pulse or heartbeat. What conclusion must we now come to considering that most of the observations indicate the lion is dead but we have one that indicates the opposite is true?

We must study this further because we have just saw evidence that refutes the idea that if a lion moves its legs it must be alive. We know now, because of someone else's experimentation and observation, that we have witnessed what is known as rigor mortis. This phenomenon occurs over time after death. The muscles contract and become rigid, or stiff, due to chemical changes that occur in tissues after normal life functions have ceased.

This scenario illustrates the flexibility of science. We have encountered an instance where something that has been supported over and over is refuted. Science then begins to explore this strange new phenomenon in depth and begins experiments that test these new findings. We will not carry this scenario further at this time to keep you from getting too squeamish. We can see from this that science is ever-changing because it is subject to tests and observations from anyone, anywhere, and at any time.

The ability to reproduce any supportive or refuting observations and experiments must be preserved through the careful recording of all variables present. This allows others to attempt as close as possible a reproduction of the original observation or experiment and achieve the same results. This is exactly how progress is made over time that sometimes benefits and sometimes detriments all the living creatures of the planet Earth.

Faith and Science

How does all this relate to the discussion at hand? To be able to look objectively at both sides of this issue both sides must be able to agree on certain definitions of the key terms that we will use in this discussion. Let us go back to the first key term: Faith. Faith is the belief in something without any tangible evidence that it exists. Some at this point may argue that they have had personal experiences or have even seen a miracle that proves to them that God exists. This is not a discussion over whether or not God exists. I myself believe that God exists and my belief is based upon my faith.

Let us now apply our nature of science definitions to faith. The notion that they do not coincide is the first thing that is noticed. One can say that they have had a personal event that proved to them beyond a shadow of a doubt that God exists. I have had such an experience myself. Is this experience reproducible? Can others simulate like conditions and achieve the same result every time or even nearly every time? The obvious answer here is no. At this point we can agree that faith has little to do with science or the nature of science. That is why we have a defining term – faith – for something that we believe without any supportive evidence.

To summarize we can agree that the nature of science is that it is made up of ideas that, through observations and/or experimentation, can be supported or refuted. If an idea is always supported through observation and/or experimentation we can call that idea a theory. Theories are always subject to change and modification or additions and deletions through the refuting and reproducible results of observations and experiments of others. I have introduced a new term here in the form of reproducibility. Others in different locations must be able to simulate the conditions under which you made your observations and or experiments and expect to achieve essentially the same results. In other words part of the nature of science is that any results of experiments or observations must be reproducible.

Science and the Bible in Agreement

The Bible does not always disagree with science as many instances can be found where science even supports what is found in the Bible. Let us look at just one: the Bible seems to converge with the idea of survival of the fittest on some points. Just one would be the issue of faith. The Bible teaches us that we must have faith in God. It also indicates that the more faith one has the greater their place will be in their relationship with God. Could we equate faith with strength and fitness?

I would say that one whose faith is great will easily endure the trials and tribulations placed upon them throughout life while those weak in faith will have a tougher time persevering. In a sense this lies in line with the survival of the fittest idea. We can also equate sin with disease. One who is consistently covered up in sin and never repenting nor asking for forgiveness can be said to be diseased to such a great extent as to impact their fitness. When you compare this to one who strives to walk in the way of our Lord, always repenting and asking for forgiveness, they would have a much greater level of fitness than the previously described person.

Another of these important terms that must be understood is the term evolution. Not to be confused with the theory, we only want to define the word itself at this time. Evolution is simply change. Human babies can be said to evolve into toddlers, then evolve into teenagers, then evolve into adults. The most prominent changes or evolutions that are observed in this example are the size of the human; it increases over time with age.

There are other less noticeable changes, or evolutions, such as the development of the mind, intellect, and a person's ability that all evolve over time with experience. We will examine the concept of evolution in much greater detail later. At present, we only need to understand that evolution is simply defined as change.

Another term we will use is adaptation. Again we can define this term using a human example. What happens when we get too hot or too cold? We either seek shade and maybe something cold to drink in response to the heat or we increase the density or layers of our clothing in response to the cold. In other words, we adapt to our environment. Change in temperature is just one example of events that occur within relatively short time periods. We will later examine adaptation as it relates to long-term changes that occur over many generations.

Now that we have some basic terminology down, we will take an in depth look at both sides of the issue. We will first look at them individually and break them down into easily examinable components. To begin our exploration and examination, we will examine the Creation event, as found in Genesis chapter one of the Bible, first.

Chapter II

Creation

The Biblical Account of Creation

The divinely inspired Biblical account of creation begins with the first verse of the first Book of the Bible, Genesis. For the purposes of this discussion I shall use the King James Version of the Holy Bible, which is one of the most commonly used throughout modern Christianity. Although there are several versions of the Holy Bible, among the many King James Versions, I have found little difference in the alteration of the intended meaning, if any at all.

The six days that contain the work of creation take place from verse three to verse thirty-one. There are a few important items that require attention while reading this the first Chapter of Genesis. Some may have noted that I observe the start of the first day of creation as verse three. Verse one and two read:

> 1 "In the beginning God created the heaven and the earth."

> 2 "And the earth was without form, and void; and darkness *was* upon the face of the deep. And the spirit of God moved upon the face of the waters."

There are no distinctive measurements of time to be found within these two verses with the exception of "In the beginning..." denoting that this is when the heaven and earth were created. The second verse points out the fact that there was no form, or shape, to the earth. We are also told that there was darkness in some place where there was depth and that water was present. How long this period of time lasted is not indicated at all but we know that there is a heaven and an earth which has no definable form.

First we will examine the words "...without form, and void..." We all know that the earth is roughly spherical in shape with many mountains and areas of depth as well. To be more specific, the tallest peak, Mount Everest, is nearly five and a half miles tall and the greatest depth below sea level, the Marianas Trench, is slightly more than six and three quarter miles deep. If it were possible to remove all the water and vegetation from the earth a rough sphere with many spikes and deep depressions would be observed. Defining void is relatively simple. According to Webster's void is defined as "not occupied or inhabited" and "containing nothing". In other words, there was no life upon or within the earth which, at this time, has no definable form.

Modern theory suggests that as our solar system was created the planets, including earth, were bands of matter orbiting the sun not unlike a series of asteroid belts. These bands of matter, due to the velocities of their orbits and the proximity of their constituents, eventually collected together due to gravity and formed the planets. This would certainly explain the "...without form..." statement. In addition we have a case where science supports what is found in the Bible. We also have the inverse: the Bible supports what science has theorized.

What we have just illustrated is the application of both a Biblical and scientific test to a theory. We have just reinforced the scientific view by finding support for it within the Bible. We have also found an instance of where what is found in the Bible is reinforced by what we have found in science. That is the most exciting premise of this book. If a Biblical and scientific test can be applied to any concept then we can have support or refute of any of the controversial issues that plague our society today and in the future.

Origins of the Bible

To better grasp these ideas let us look to the origins of the Bible. It is believed that the first words of the Old Testament were written by God Himself in the form of the Ten Commandments that were provided to Moses over 3,500 years ago. The Old Testament was completed sometime around 500 B.C. or over 2,500 years ago.

Humans of that time had an entirely different view of the earth and other humans than we do today. These humans also had entirely different terms and ideas which they used to put events into a context that would be understandable to them selves and others. For the next part of our examination, let us try to gain an understanding of the time in which the Bible was written.

Vocabulary

It is a fairly well known fact that the human vocabulary increases with new words that come into existence on a nearly daily basis. The English language, which was used for the King James Version of the Bible, is a picture language. A picture language is one that as you listen to it spoken or read it a picture or image is formed within the mind. The Hebrew language that the Bible was originally written in is also a picture language.

William Tyndale, the first to print the Bible in English, once said that it was "...ten times easier to translate Hebrew to English than to any other language." Since both Hebrew and English are picture languages and Hebrew translates so easily to English one can be somewhat sure that the intended meaning in English has not been altered severely from the original meaning in Hebrew, if any alteration exists at all. This brings us to a key question: Did a word even exist when Genesis was written that could have described the shape of the earth in terms that people of the time could understand or comprehend?

There are 783,137 words in the King James Version of the Bible that can be defined as English, Latin, Greek, or Hebrew. The word sphere is not found once and the closest similar word, ball, is found only once in Isaiah twenty-two verse eighteen and not in the context of describing a shape. Many words, in fact, are not to be found anywhere in the Bible which presents two possibilities: the words did not exist at all at the time or God, through Divine Inspiration, chose not to use them. One would have to assume that God truly wants us to understand the Bible and would therefore use every means available, including word usage, to convey the clearest meaning possible.

You can also look to the fact that for centuries, even during the time of Christopher Columbus, humans believed that the earth was a flat disc. They believed this to the point that they feared falling off the edge if they were to approach it. There is only one conclusion one can come to: the Bible could not have been intended as a collection of riddles and hidden messages designed to be figured out by some futuristic code breakers.

The Bible was Divinely Inspired by God to be easily understood by those He inspired to write it and those living at the time it was written. In order to be written and understood by humans the vocabulary would have to consist solely of words that were available and in usage over 2,500 years ago. By this same line of reasoning it can be safely deduced that the concepts introduced would also have to fall within the realm of the then current understanding.

One could argue the description of supernatural events within the Bible would be beyond the human intellect's grasp of the time. While it is true that there are numerous supernatural events described within scripture, they are always defined using terms that could be comprehended and therefore imaged in the mind. Once again, we reinforce the concept of the picture language. If you start using the terms 'asteroid' 'orbit' and 'gravitational force' what mind could have grasped, let alone imagine a picture, of these concepts over 2,500 years ago?

The Times

The Bible was written in a time of kings and empires. There are accounts of the belief in race superiority where one would enslave the other. These accounts are mentioned in the Bible with the Egyptians and the Israelites being the most detailed account. After having been justified by the Bible, according to Genesis one verse twenty-six, that humans were to

> 26 "...have dominion over...all the earth, and over every creeping thing that creepeth upon the earth."

How long would the theories like evolution – or any notion that was beyond the grasp of the human intellect of the time – lasted in those days? Any book that contained such notions, not to mention the person who wrote them, would have been banned and destroyed by the rulers of the day.

Take, for example, Einstein's Theory of Relativity. We will not go further than to mention it as an example as it is far too complex and beyond the scope of this work to discuss in any detail. Would the people of the times, let alone their rulers, have the capacity to imagine and embrace this concept? Would they even read beyond these opening verses to get to the 'meat' of the scriptures if the Bible had started with complex and incomprehensible writings? I think it would have been more likely that they would have banned and burned the book as well as its writers.

We find in Genesis Chapter one verse twenty-six that God desires his human creation to

> 26 "...have dominion over the fish of the sea, and over the fowl of the air, and over the cattle, and over all the earth, and over every creeping thing that creepeth upon the earth."

This idea would have been extremely popular with people and especially the rulers of the time who would have received their justification from this verse.

The rulers and kings throughout time would have ensured the survival and distribution of such writings that seemingly justified their rule and, as King James did, have them translated into his subjects' language. This certainly would have been anticipated and therefore planned by God so that His word would survive perpetually and be spread to the widest audience possible through those having the best capability to do so throughout time.

The Parable

Within the Bible the parable is used extensively, especially by Jesus when teaching others. The Cambridge Dictionary defines parable as a "term that translates the Hebrew word "mashal"-a term denoting a metaphor, or an enigmatic saying or an analogy." At this point we can agree that the parable is used within the text of the Bible and see that the purpose of these analogies was to provide a clear and easy to understand method of explaining a concept to the people of the times so that it could be imaged and accepted within the mind. The Bible was Divinely Inspired meaning that God put the words down through the people who wrote them. Following this reasoning it can be deduced that God does teach in parables when necessary to convey a concept that may be beyond the reach of human intellect for whatever reason. Is it too much of a stretch to believe that God may have been teaching in parable the account of creation in order to convey His message in a way that could be comprehended by humans of the time?

The Bible and Continental Drift Theory

According to Genesis one verse nine there was only water upon the surface of the earth until the third day when the water was "...gathered together unto one place..." to allow the dry land to appear. If you take this verse literally, you must agree that there was only one land mass at some point in time. If the water were gathered into a single place, meaning one gigantic ocean, then there could be only one land mass.

Today there are seven major continents spread over earth indicating that over a period of time the original one continent broke into the seven we observe today. There is a theory of continental drift that puts forth the idea of a single land mass that slowly broke into the continents as we know them today. Theorists have given a name to the single continent that existed over 250 million years ago: Pangaea.

The Theory of Continental Drift states that the earth is made up of twelve plates that are slowly drifting away from, towards, and against each other at the rate of one half to four inches per year. The ones that grind against one another, such as at the San Andreas Fault in California, can cause earthquakes. According to the rate of movement of these plates scientists have estimated that the singular continent, as described in Genesis, would have last existed as a single continent about 250 million years ago.

Scientists have been measuring the movement of the continental plates for decades and detected movement that ranges between one-half and four inches each year. This movement of the plates is evidence to support that the Bible states exactly what did occur early in earth's history and we therefore have a case of science backing what is stated in the Bible. We also, once again, find the inverse to be true: the Bible supports a scientific theory.

One piece of that supporting evidence places North America farther south and east of where it is today. While North America was a part of the single super-continent, it was located on the equator giving the area a tropical climate. There have been fossils located in the fairly cold climates of North Dakota and Canada of tropical plants and animals found today only in equatorial regions. This is even more support for this idea and therefore support for what is written in the Bible.

Another scientific support of the single continent first described in Genesis Chapter one verse nine is the existence of placental mammals and non-placental marsupials and their distribution upon earth. Placental mammals are those mammals that develop within a placenta in the female of a species. Humans are an example as are cattle, horses, lions and the like. Non-placental marsupials, while they are still mammals also, differ in their embryonic development. Non-placental creatures are born extremely underdeveloped and usually spend the first part of their lives within a pouch. The kangaroo is one example of a non-placental creature.

Upon the single land mass the placental and non-placental mammals coexisted, utilizing the plentiful resources to be found on this gigantic continent, according to the fossil record of many continents at the time of the single continent. When the continents started to drift apart some of the non-placental marsupials migrated to the Australian continent over a land bridge that connected it to South America. These non-placental marsupials, like the kangaroo and koalas, thrived on their own in Australia without the presence of placental marsupials. Those non-placental mammals remaining on the South American land mass, with the single exception of the opossum, were extinct in short order due to competition with the placental mammals for the now more limited resources. This is also an example of natural selection, or survival of the fittest, as we shall see later.

Chapter III

Evolution

The Primordial Slime

The first picture that comes to mind when we think of evolution is this puddle of primordial soup or slime that all life was supposed to originate from when this creature crawled out of it. The Theory of Evolution is actually a group of ideas based, for the most part, on supporting observations of Charles Darwin and a few ideas of others that were combined to form what we know of as the theory.

To properly examine the theory we first have to break it down into its individual components and carefully study them one by one. Let us first briefly examine the times and the man himself. There were many naturalists in Darwin's time and theories were being presented almost daily that sometimes contradicted one another. This was an era of rapid discovery where many sound theories were presented as well as many unsound ones.

Charles Darwin

Charles Darwin, who is said to be the 'father' of evolution, was a scientist, and a relatively wealthy man by the standards of his time. He was presented with the opportunity to accompany a voyage aboard the HMS Beagle as the ship's naturalist in 1831. The voyage was to carry them to South America from his home in England for a two-year survey (they actually did not return for almost five years). It was on this voyage that Darwin observed many of the phenomena that led to his formulation of the Theory of Evolution.

Throughout the voyage Darwin made many inland excursions collecting rocks, fossils, insects, plants, and animal skins that he sent home for further study upon his return. Many of Darwin's collected specimens were also sent to scientists at various institutions for further study to be completed before his return. He amassed all this information upon his returned and began an intensive period of study.

It was Darwin's observations and studies that led to the formulation of his theories. As we will soon see, he made many assumptions or hypotheses to attempt to explain various phenomena that he observed and, for the most part, these hypotheses were valid but only when taken in context of the time. Science was still in what could be called the 'dark ages' when taken in the context of modern times.

Upon close examination of Darwin's work as he wrote it you will see that there is very little mention of humans and their place in his theories. It must be kept in mind that

Darwin mentioned absolutely nothing of primordial soup or slime in his writings. These ideas, along with most of the general perception of the Theory of Evolution, came about during the last several decades as science, the media, and the general public added their own 'modern' translations to the theory. What we perceive today about evolution differs greatly from Darwin's work.

While it is tedious reading for the general public Darwin's "Origin of Species" is published today by Gramercy press and can be found nearly everywhere. It is among the most misrepresented and misinterpreted works of all time. Most know of Darwin and then his theory and they immediately think of 'man descended from apes'. Nowhere within the book's six hundred pages is there any mention of where humans may have descended from. The root of the wide chasm between the sides in the debate is a gross misunderstanding directly caused by gross misinterpretation and incorrect translation and assumptions. No, I am not saying that Darwin was correct about everything. I am stating that most of what the general population attributes to Darwin is incorrect.

We will now break these ideas down into their individual components for study on an individual basis and, at the same time, look at some other theories that are pertinent to our discussion. Our discussion will be slightly widened to include theories that in any way relate to the Biblical account of Creation. I do this broadening to accomplish one of the purposes of this work: to bring science together with the Biblical account of Creation so that the Bible will be found to support science and science will be found to support the Bible.

As I have done throughout this book we will utilize the Biblical and scientific tests for each idea or theory. The scientific test will explore the observations and/or experiments that are reproducible that can be found to either support or refute any particular idea. The Biblical test will either find direct support of the idea within the Bible or find that what is presented in the idea does not diverge significantly from what is found within the Bible. I will also illustrate the concept of universal understanding. It is only through a delivery of universal understanding that either side can have any hope of a universal acceptance of an idea or theory.

Chapter IV

Persistent Theories

The persistent theories presented here contain not only the individual components of Charles Darwin's Theory of Evolution but other theories that either were formulated by others before or during his time as well as more recent theories that support or reinforce the individual components.

The Theory of Natural Selection

The Theory of Natural Selection is the easiest to explain and understand. This theory is the basis for the 'survival of the fittest' idea. Natural Selection is the process that finds that the strongest, or more fit, of a species is much more likely to survive and reproduce no matter what environmental challenges it faces. The weaker members of a species, unable to cope with changes in their environment, are less likely to survive to reproduce. This process allows the original creation's descendents to be ever-improving, or evolving, as time passes hence prohibiting undesirable traits from being passed down.

Let us look at an analogy to help us understand this idea: the lion. We are observing two lions in the wild and notice that one of them is apparently normal and the other has a slightly deformed rear paw. There are no other apparent differences between the two. The lion with the abnormal paw can run, but not quite as fast or for the same duration that the other lion can. We know this because we observe them tracking, chasing, and capturing their prey. We observe that the lame lion can only catch slower or injured zebra while the stronger lion has no trouble bringing down any that he desires.

The differences really become apparent when we notice that the prey's population has declined (reasons for this are not pertinent to this discussion) and so with fewer available targets the lions are taking longer to find and capture their prey. We note that now the abnormal lion is getting a lower percentage of the overall prey captured because he becomes tired out quicker than the other lion. This only leads to the lame lion becoming weaker and more susceptible to disease as time passes. Eventually the lame lion succumbs to starvation and disease that has run rampant throughout his weakening body.

This is an illustration of survival of the fittest in a nutshell. The lion without the abnormality was better suited to adapt to the decline in the population of his prey than the other and therefore survived to pass on his superior traits to any offspring. This process is repeated again and again throughout all the generations since the species was created. This same scenario can be expanded to the prey of the lion by imagining that the weaker and therefore slower zebra will be primary targets for the lions. Again we have an illustration of survival of the fittest. We could expand this scenario on any species upon the earth today or those of the past.

The Theory of Natural Selection simply states the idea that through the occurrence of natural events, without intervention – Divine, human, or otherwise – the weaker members of a species will typically face demise prior to reaching an opportunity to reproduce. This natural check on the ability to reproduce not only helps to control the population, but also helps to ensure that any undesirable traits are not perpetuated, or passed on to successive generations.

The inverse of the above also holds true: the stronger members of a species are more likely to survive and reproduce passing on their traits to new generations. This ensures that those traits that are beneficial to the organism's survival are maintained perpetually throughout the generations. This is yet another support of the idea that these built-in controls and measures required an intentional and intelligent designer.

To further our illustration we can look at what may have happened if we could have intervened on the lame lion's behalf. Perhaps he could have been captured, taken to a zoo, and fed meat till he lived a ripe old age. We might have even provided a mate and allowed reproduction to occur. Would the deformity have persisted into the offspring? There is no simple way to tell without going into details that are beyond the scope of this book.

If we were to intervene on the lame lion's behalf the only thing that we can be assured of is that, over time and generations, we would gradually reduce those beneficial traits necessary for the lion's survival in the wild. We will, in effect, create an entirely different creature over time that would be entirely dependent upon humans to provide for its needs. This would, in turn, create a less capable creature that was unable to carry out the purpose for which it was created for in the first place.

The point to remember from the above illustration is that without any intervention nature would take its course. The lame lion would most probably face certain demise prior to maturing and gaining the opportunity to reproduce and thereby eliminate the possibility of passing on a detrimental trait to any offspring. This is by no means cruel as far as the intent of the designer is concerned. The reason the lion was lame could have been due to an unexplained alteration to its DNA brought about by an environmental event. If such an alteration to DNA did occur, it is fairly certain that, if permitted to reproduce, the deformity would persist in any offspring.

Extinct Species

A look at extinct species also gives us a clear illustration of the Theory of Natural Selection from a different perspective. There are many reasons why species have become extinct. Some species are extinct due to a change in the environment that no longer supports the particular species as happened during the Ice Age when it became too cold for many species to survive. Usually those changes that cause a mass extinction event are ones that occur rapidly and give a large proportion of species little or no time to adapt to the changing environmental conditions.

Extinction has also been initiated due to actions of humankind. Irresponsible actions such as polluting the environment, hunting endangered species, or the destruction of natural habitat have resulted in the extinction of many species. Frequently extinction is caused in nature by natural selection such as when a new species is introduced to an area where it competes with another species for resources and one or the other proves to be the weaker of the two and succumbs.

The Haitian Solenodon, an insectivore (a creature that eats only insects), weighing around two pounds, was determined to be functionally extinct during the 1960's. Native only to the island of Haiti, the Haitian Solenodon hid through the day and searched for insects during the night for sustenance.

The extinction of the Haitian Solenodon began when humans brought along their dogs and cats to the island and, as in any society with domesticated dogs and cats, some of them were abandoned and became feral (wild). These feral dogs and cats being carnivorous, or meat eating, found the Haitian Solenodon to be an easy target. The Haitian Solenodon, not having time to adapt to this new predator, quickly succumbed and ceased to exist altogether.

The Haitian Solenodon is by no means the perfect example since we have the component of humans introducing the dogs and cats and permitting some to become feral. It should also be understood that the entire chain of events that led to the Haitian Solenodon's extinction was unintentional, as far as human intent was concerned. Just the same, we have one species proving to be the stronger and driving the other to extinction. Again, all we have is more evidence to add to our body of support that part of God's intelligent design is the population check of Natural Selection.

The other intentional and planned purpose of natural selection is to help prevent undesirable traits from being passed down to offspring. At the same time the inverse is also true: natural selection ensures that the desirable traits are passed on which allow for the creature's ability to adapt to changes in the environment or resources and permits an improvement in the creature over time. God surely allowed for Natural Selection with a purpose: He wanted His creation to survive perpetually in the ever-changing environmental conditions of the earth.

The Theory of Variation of Species: Human Induced

The Theory of Variation of Species is a little more complicated than natural selection but to illustrate this theory more accurately, I will take a more recent example that is easily understood. Our family has a Boston Terrier for our indoor pet, it really isn't relevant, but his name is ElRoy. When I started searching for a suitable pet I performed many hours of research into temperament, how they are with children, and so on. I came upon some interesting histories of the different breeds of dogs.

Everyone is somewhat familiar with the fact that all the breeds of domesticated dogs were 'created' by humans through interbreeding and selective breeding to eliminate the undesirable traits and amplify the desirable ones. A few examples of this would be the Australian Sheep dog which was bred for sheep herding and the Fox Terrier, which was bred specifically for aiding fox hunters when their quarry disappeared down a fox hole.

Boston Terriers originated in Boston, MA in the 1800's from breeding the English Bulldog with the now extinct English Terrier. The original intent was far from a noble one: to create a stout fighting dog with an abundance of energy. At any rate we ended up with ElRoy who sleeps a lot and gets so exited he can almost knock a teen off their feet when he jumps on them.

The English Bulldog, a very strong and stout breed, is also known for being extremely lazy. The English Terrier, like most terrier breeds, was an extremely energetic and flexible type. These two breeds, when interbred by humans, resulted in the extremely strong and energetic Boston Terrier.

While God did create the original predecessors to the wide variety of domesticated dogs we have today His intention was clear: a diverse variety of creatures will be formed from the originals to allow for changes in environment, or, as in the case above, changes in human needs or desires. This allows for His Creations to be of the perpetual kind, or always replenishing, as He directed them to do during the Creation event. These creatures had the ultimate design within them from the start, placed there by God Himself to allow for any changes, whether they are caused by nature or humans themselves.

Naturally Occurring Variation of Species

We have explained variation of species from the human-caused perspective with ElRoy, but let us now look at a naturally occurring example. We will examine one of the phenomena that Charles Darwin observed when we take the example of the male Peacock. It has been observed that male Peacocks have large beautiful displays of colorful feathers on their posteriors which they use to attract potential mates. In the presence of females during a mating opportunity they flare these feathers out to cover the greatest area possible and strut about, apparently to impress the females of the species and initiate a mating opportunity.

It has further been observed that a female, when presented with a choice of two males, will choose the male with the largest display. We know, from direct observation over many years, that male Peacock feather displays grow larger over many generations – never smaller. In other words, larger displays in male Peacocks are favored over smaller displays.

This is an example of a beneficial trait being passed down and amplified over time due to the fact that the males with the largest displays are presented with more mating opportunities than males with smaller displays. Due to the inverse we have an undesirable trait, small displays, being phased out due to the reduced mating opportunities presented to males with lesser displays.

In the above example the logical argument was presented that as feather displays grow would the increased weight or mass of the males make it more difficult for these males to escape predators? Would this not make large displays a hindrance rather than a benefit and cause the smaller displays to be favored? When you look at the population overall and see that there are far more male peacocks with larger displays than there are those with smaller displays then the only obvious answer is no.

While it may be true that male peacocks are easier prey due to the increased mass of a larger display, the benefit of increased mating opportunities outweighs the detriment of increased mass in the case of the Peacock. The sole support for this theory is that there are more male peacocks with large displays than there are males with smaller displays indicating that the trait for larger displays is passed on to offspring more often. This is another example of observation in action which is a cornerstone of science.

There is also an example that has a mixture of these various theories illustrated. Just prior to the industrial revolution in Great Britain there existed a species of moths. The Peppered moths had a built-in camouflage system that served to protect them from their main predator, birds. They varied in color with some being dark and others being significantly lighter. These moths would find refuge during the day by roosting on the bark surfaces of trees that closely matched their particular coloration. Since, in the area studied, there were more numerous lighter colored trees than darker colored trees there were more light colored moths than dark ones.

As the industrial revolution began along came pollution in the form of dark ash and soot that began to cover everything – including the vegetation. Over a period of years, it was noted that the light colored moths mysteriously disappeared as none could be found in the areas affected by the pollution. The darker colored moths were found in an abundance never before observed. The reasons: natural selection and variation of species. The lighter colored moths were no longer protected by their built-in camouflage due to the soot deposits on all the trees which made the bark significantly darker. With no refuge they were easily spotted targets for the birds that rapidly devoured them.

The lighter colored moths could not live to maturity to reproduce whereas the dark colored moths proliferated now that all the trees' color favored their camouflage. This example illustrates perfectly how a negative trait can be phased out of a species while, at the same time, a positive trait can be amplified, passed on to offspring, and even lead to a proliferation of the species. While this event occurred very rapidly, most similar examples occur over a much greater expanse of time.

Variation of Species: Built-in Population Controls

Another component of variation of species that directly relates to the Bible is its built-in natural method of population control. Although God directed his creations to multiply and either fill the waters or the earth He knew that unchecked reproduction would eventually result in depletion of resources and His Creation would eventually destroy itself.

The above idea can best be illustrated by using a pair of great whales. Let us assume that each mating pair of whales are capable of producing six offspring during their lifetimes. Each pair of their offspring has the same capability and so on. You will have the original two in the first generation, six in the second, eighteen in the third, and so on till there would be over one million members of the thirteenth generation. I am only counting those born and not any surviving whales from previous generations.

You can see that just one species alone, growing in numbers and unchecked by any controls, would exponentially grow until the resources of the earth could no longer support them. The additive effect of all the species that God created growing in numbers exponentially would very quickly result in the self-destruction of all God's creatures due to a rapid depletion of resources.

Population control in the form of variation of species and natural selection is now shown to be a part of God's intelligent design from the beginning. These ideas are supported by both scientific and Biblical principals. Nothing within these ideas is refuted by science. To be even surer, nothing within these ideas refutes what is within the Bible, in fact, they support what is in the Bible.

The Theory of Use and Disuse: Domesticated Species

The Theory of Use and Disuse states that the less some item of a body is used, the less pronounced it becomes, until it eventually disappears. The inverse is also true, the more often an item is used, the more pronounced it becomes. Darwin gives the example of domesticated cattle. Some of the older breeds have ears that droop rather than stand up.

The idea above is that, over many generations, these domesticated breeds were startled less and less by their familiar surroundings. One of the reflexes of being startled is that the ears prick up and turn toward the direction of the sound that startled them in the first place. As the cattle became more familiar with their limited surroundings of the confined barn yard the startle reflex of the ears occurred less and less. This reduction and eventual lack of use eventually led to the ear muscles and cartilaginous structures to relax and droop over many generations.

Another example that Darwin gives is the difference between wild and domesticated ducks. When comparing the leg bones of the wild and domesticated ducks it is found that the domesticated duck has heavier and stouter leg bones than the wild one. One of the key differences between the two is that while the wild duck flies frequently, the domestic duck walks most of the time. The domestic duck, having to walk most of the time and over many generations, has developed heavier leg bones in response to the increased use of the legs. None of this evidence refutes anything that is found in the Bible and it is all scientifically sound. Therefore we can accept the theory of use and disuse as Biblically and scientifically sound.

The Theories of Mass Extinction Events

Theories abound on the subject of mass extinction events. The most studied extinction event concerns the one that affected the demise of the dinosaurs. It is generally agreed that there were at least six mass extinction events during the earth's history. It is theorized that the event that ended the reign of the dinosaurs occurred approximately 65 million years ago when a giant meteor struck the earth off the Yucatan Peninsula in the Gulf of Mexico. It is surmised that the impact and its effects upon the climate caused the destruction of nearly 85 percent of all the species in existence at the time. While the dinosaurs were entirely wiped out, other creatures such as mammals, some amphibians like the crocodile, lizards, and the birds somehow survived this mass extinction event and even prospered in what had to be a very harsh environment.

Oil and the Dinosaurs

Perhaps the following illustration will aid in your understanding of how this could possibly be a part of God's plan. The world's crude oil resources are found beneath the ground at various locations and depths throughout the world. Crude oil, in its natural form, is the result of the decay of primitive biological material, including plants and animals, beneath the ground with some minerals from the earth itself mixed in. Mass extinction events are the only explanation for the existence of these vast crude oil deposits. Further support of this idea comes from the fact that the crude oil deposits are found at varying depths, indicating a number of mass extinction events.

To carry this line of reasoning even further the Bible does give us an idea of the time elapsed between the creation of man and the birth of Christ. The extensive genealogies within the Old and New Testaments of the Bible give us a fairly precise timeline. When you add the rest of time since then until now, you have an idea of how long humans have been present upon the earth. We have just found more Biblical and scientific support of the idea that the only place for all this unaccounted for time is during the Creation event.

If you follow these ideas further you can agree that all these events had to be perfectly orchestrated at the right times to produce the resources that humans need today. There are so many humans on the earth at present that we could not survive without our natural resources to provide food, heat, energy, and fuel.

What are you saying, Glenn, that God gave us oil? Absolutely! How else – other than a Divine and intelligent design – would all the events and ingredients come together so perfectly? We had to have all these primitive creatures, among them the dinosaurs, along with the vegetation present at the time. In addition, all of these plants and animals had to be spread throughout the earth, as God had directed them to in Genesis.

The key to understanding this idea is to keep in mind that oil is found, in various sized deposits and at varying depths, all around the world. There are vast oil fields in the Middle East, the south-central United States, and even beneath the waters, as in the Gulf of Mexico. All these creatures and plants had to be destroyed, in one cataclysmic event, leaving mammals, along with a few other species unscathed. Since there are some very sizable pockets of oil in locations throughout the earth then one could assume that the mass extinction event that triggered the demise of the dinosaurs destroyed them fairly rapidly and without much of a warning.

Think about this for a moment. Mammals are one of our primary sources of food, created and saved for us by what could only have been a miracle of God. These creatures are quite amazing to have survived the elimination of eighty-five percent of all species on the earth. It is not that surprising that our Divine Creator also chose to model us along the same lines – we are mammals too. No, no, no! I am not suggesting that we are descended from other mammals, as Darwin erroneously assumed. I am simply stating that God had already had millions of years of experience observing His various creations.

When God decided to create man He knew that dinosaurs would not get along too well with us. He also couldn't place man on a desolate earth with no life at all. God knew that mammals, such as cattle, would be a fine source of sustenance for humans. He also, through His study of the mammals, incorporated many of the same characteristics into His humans. Mammalian reproduction, placental development, and the way the female of the species nurses her young are all traits that we share in common with the mammals that God placed here before the humans. "...and God saw that it was good."

God also knew that His destruction of the other species would one day be transformed and discovered to serve His human creation in the form of oil. Oil is one example of a major discovery that drove the progress of humans and, at the appropriate time, we were permitted to find this wonderful resource. A Divine and intelligent plan is the only way that all these events could have been orchestrated.

The evidence is hereby presented that these events were yet another part of God's intelligent design during the creation event to make room for other creatures upon the earth and provide us with the resources necessary to support an ever-growing population. God actually experimented with, tested, and observed His creation over a vast period of time. All His creatures had the built-in ability to adapt and improve to better meet the demands of an ever-changing environment.

Chapter V

The Big Bang

In the Beginning

Science and the Bible do agree directly on one point here: There was a time when there was nothing and then... A Beginning. While the Big Bang Theory is not pertinent to our study of Creation as it pertains to Evolution, it is a debated event of Creation. I have, therefore, decided to devote a chapter to it due to the importance of one of the purposes of this book: bringing science and the Bible together as it pertains to Creation. The theory does not fit well into the previous chapter as it has little to do with evolutionary theory.

Expansion From a Single Point

Supported by science, as first observed by astronomers in the late 1920's and many others since, is the idea that the universe is constantly expanding. This premise has to allow for the idea that the universe was once compacted – to a single point or singularity. At one time, before time even existed, nothing at all existed. This single point appeared and gradually expanded to the universe as we know it today. If we go back to Genesis chapter one verses one and two we find the Biblical support for this idea:

> 1 "In the beginning God created the heaven and the earth."
>
> 2 "And the earth was without form, and void; and darkness was upon the face of the deep. And the spirit of God moved upon the face of the waters."

There are no accounts or descriptions of time to be found here. Any attempt to apply time restraints upon these verses could result in guesses of minutes, hours, days, years, millions, or billions of years. Science and the Bible therefore agree that there was a singular event that marked the creation of the universe. In other words, there was a beginning and it could have started thousands, millions, or billions of years ago.

There was probably no explosion, as the Big Bang would bring to mind, but rather an expansion from nothing to something. In verse sixteen of Genesis chapter one we find God creating:

> "...the greater light to rule the day, and the lesser light to rule the night: *he made* the stars also." (Italics in original).

Here is the specific account of God creating the sun, the earth's moon, and the stars. We also know that the other planets in our solar system exist and, therefore, their very existence tells us that they had to be a part of this creation event. Religion and science both do agree on this point – BUT – the planets are not mentioned in Genesis chapter one. Are we adding to the meaning of the Bible if we concede the point of the creation of the planets? This is just another illustration of the repeated support that we have found within the Bible that the account of Creation is not to be taken in an entirely literal manner.

Keep in mind this one point: I have never, throughout this book, disputed a single word contained within the Bible. I have only proposed a very plausible interpretation that I have shown to be supported by science and what is written in the Bible. The other amazing point to remember is that not only have I shown support *from* the Bible, I have provided support *for* the Bible.

Birth of a Solar System

Science supports the notion that as our solar system developed there were rings of rock, dust, and other matter orbiting around the sun. This definitively explains the "...without form..." statements without a doubt. It is believed that these loose rings gathered together within their respective orbits and formed the solid masses that we know as planets.

The joining together of the various materials to form the planets had to occur over what had to be a vast period of time. That period of time over which this occurred is not, and never will be, precisely defined as this definitely falls under the realm of 'only God knows'. Once again, we have another example of science supporting the Bible and the Bible supporting science.

The Big Balloon

To reinforce and illustrate the correction to the most common misconception about the Big Bang Theory we will use a balloon to illustrate the concept. Remember, at first there was nothing. Then at a single point in time and at a certain place there was something. Let us think of this something as an uninflated very tiny balloon. Everything that is in the universe today as we know it is inside this balloon. The balloon is ever expanding and will never burst – we hope. All the galaxies, solar systems, and the unimaginable vastness of space are contained within it.

We do not know where this balloon is or where it came from. We do know that objects in the universe are continually moving away from each other at speeds relative to their distance one to another. We do know that the universe follows certain laws as far as its behavior is concerned from the perspective of physics. These laws could only have come about with intent and from an intelligent designer – God.

Chapter VI

The Theory of Human Evolution

The Controversy: Questions

Human evolution is such a controversial subject that even Darwin would not elaborate on it for more than a sentence in his work. The questions have remained the same since humans could express themselves and these same questions remain unanswered to this day. Where did we come from? Why there are different races of people and what are their origins? What caused the differences between individuals on the earth?

The above are questions that have been central to human curiosity. Perhaps the most vexing question would be "Why are we here?" While I have pondered that very question just as anyone else has, I will stick to the questions I posed before that – in this book anyhow. Once again, I will restate my primary purpose for creating this work: To bring the sides together that are divided in debate.

The subject of human evolution is such a controversial one that I am sure the conclusions regarding it that I present here have potential to create debate. The purpose of this book is not to create debate but to bring the sides together in a universally understood and acceptable theory. It is my sincere hope that any debate created be productively aimed at bringing the sides together rather than divide them further.

Human Evolution: Environmentally Induced Change

Human evolution, or change, is mostly based upon environmental conditions present wherever humans are found upon the earth. The first human creations, Adam and Eve, are described as having been formed near the Garden of Eden. Based upon the description in the Bible most experts agree that Eden would have been located in the Middle East near where the Tigris and Euphrates rivers meet. The record of human history also agrees that this is the general area where human civilization began. The Middle East can be considered an intermediate area as far as climate is concerned. There exist the extremes of the Northern Hemisphere as in the Polar North to the extremes of the sub-tropical arid areas such as Africa.

Earth's Climate and Solar Radiation

Let us look for a moment at the earth's climate and how the sun affects it. The earth rotates about its axis as it orbits the sun. Each rotation faces us toward the sun in the day and away from it at night. The equatorial regions of the earth are subjected to the direct straight-line exposure to the sun's radiation. As one moves away, to the north or south, from the equator the sun's radiation becomes angled to an ever-increasing degree the further away one moves. This is what gives us our polar regions that surround the north and south poles.

The atmosphere also plays a filtering role for the sun's radiation. Without the filtering action of the atmosphere everything would be destroyed instantly by the sun's radiation. The atmosphere is of a fairly uniform thickness surrounding the earth. If you imagine the sun's radiation passing through the thickness of the atmosphere at the equator, it would pass directly through the layer and strike the earth. If you now look at his same process further north or south, the radiation now passes through the atmosphere at an angle. The distance that the radiation must pass through the filter of the atmosphere increases with the angle as you move further to the north or south. Therefore, the level of radiation reaching the earth's surface decreases as you move farther away from the equator.

In simple terms, the rotation of the earth about its axis, the orbit of it about the sun, and the tilt with which the earth's axis deviates from the vertical all affect the directness or indirectness with which the sun's radiation strikes the surface at any given point on the surface of the earth. This is what provides for the seasons and the great variety of temperatures that the earth experiences. A discussion of other weather phenomena such as storms and hurricanes is not really pertinent here so we will limit this discussion to temperature. The measurement of temperature is closely linked to the intensity of the sun's radiation at any given location on earth and that is what we are exploring.

Distribution of Humans on Earth

God directed the first pair of humans He created to "...be fruitful and multiply and fill the earth..." and designed this to be carried out through reproduction. He also created His first pair of humans with the characteristics necessary to survive in the environment in which He placed them. Although there exists no pictures or description of what Adam and Eve looked like we can get a good idea by looking at the people indigenous to the area from where they were created. Today the inhabitants of the Middle East are of an intermediate complexion between the range of light and dark-skinned people of the earth.

The climate of a location on earth is related to the angle of the sun's radiation as it strikes the earth's surface. For the purpose of illustrating this in the simplest terms possible, we will look closely at three areas of the earth. Prior to the colonization of the Western Hemisphere we find the concentration of human population to be throughout an axis running through the areas of Europe, the Mediterranean, and Africa. Let us take three specific areas that all should be familiar with: Great Britain, Spain, and Central Africa. We specifically will look at the people who are indigenous to those three areas. This simply means people that can trace ancestry to the same area in which they live for many generations.

Indigenous Peoples and Coloration

Humans indigenous to Great Britain are found to be of light complexion. This characteristic can be attributed to the environment of Great Britain as it pertains to the climate. As I pointed out earlier, the climate of an area is directly related to the angle that the sun's radiation strikes the earth. The further the sun's radiation has to travel through the atmosphere to reach the earth's surface, the lower the level of radiation and the temperature.

In the case of Great Britain the sun's radiation strikes at an angle that is greater than the angle found in Spain and the sun's radiation strikes directly and without an angle in Central Africa. To move on we find humans indigenous to Spain, the Hispanics, to have a medium complexion, and the Africans, indigenous to Central Africa, to have a dark complexion.

We know that, when looking at indigenous humans, as you move further away from the areas where the sun's radiation strikes directly to areas where the angle increases, you find the complexion of humans getting lighter and lighter – from black to brown, to tan, to white. Why do we find this phenomenon? There are numerous reasons but we will explore only a few that are pertinent to our discussion.

Human Evolution: Sequence

We could start at the present and work our way back, or start at the beginning and move forward. Since we are discussing human evolution let us explore it as it occurred. We go back to Adam and Eve in the Middle East who are probably of a medium complexion. God wanted His creations to survive where He placed them so he would have to give them some protection from the sun's radiation in the form of a pigment coloration that would shield them.

Since the sun's radiation strikes the Middle East at an angle, God gave them a complexion that would protect them. God also knew of the angle of the sun's radiation in Great Britain and the lack of an angle in Central Africa. God did direct that humans should reproduce to fill the earth so He knew that humans would one day inhabit Central Africa and Great Britain and every place in between.

We know that the sun's radiation is damaging to the skin. The degree of damage is also known to be greater in lighter-skinned humans than in dark-skinned ones. If you take a light-skinned human from Great Britain and place him or her in Central Africa without the modern convenience of sunscreen, you will observe skin damage in the form of sunburn. If we permitted continued exposure over time to this human we know that the probable outcome would be skin cancer which is not conducive to remaining alive. What I am attempting to illustrate here is that God built-in the human ability to adapt to the climates to be found in a variety of areas upon His earth. Keep in mind that changes as they relate to adaptation occur over vast periods of time and not within a generation or two.

As the descendants of Adam and Eve moved to the south and the sun's radiation became more direct these humans developed, over many generations, darker and darker skin. The inverse is also true: as the descendants moved further north the sun's radiation angle of incidence grew greater. The further north they moved the less protection from the sun's radiation they required. These humans living in the northern climates also found themselves covering up to a far greater degree than those humans of Africa in order to protect against the colder climate. Here we have the case of the medium complexion of the originals growing lighter as less protection is needed from the sun's radiation.

Scientific and Biblical Tests of the Theory

What we have just explored is a case of gradual change or evolution in humans occurring over a vast period of time for the purpose of adapting to differing climates. Let us now subject this idea to Biblical and scientific tests to see if it stands up. We will start with the scientific test. We are observing the indigenous people of Great Britain, Spain, and Central Africa. Our observations shall be limited to skin color and the angle of incidence of the sun's radiation as those items are what we are studying.

The humans indigenous to Great Britain have a very light complexion, those from Spain have a medium or tan complexion, and the Africans have dark complexions. We also see that the sun's rays strike the earth directly and without an angle in Central Africa. The angle increases in Spain and is the greatest of the three areas under study in Great Britain. We now have observations to support the idea that where the sun's radiation is direct we will find dark-skinned humans and as that angle increases we will find a gradual lightening of the skin color. Now we have the scientific support for this idea in the form of supporting observations that are reproducible.

Now we will check for Biblical support of this hypothesis. Adam and Eve were created in the area of the Middle East based upon descriptions given in the Bible. They were directed to multiply and fill the earth. The fact that we have the human population spread over the earth supports and is supported by God's direction to Adam and Eve as found in the Bible. We also find that nothing about our idea that disputes what is within the Bible. We have now passed the Biblical support tests to support this idea. Joining this Biblical support with the scientific support that we discussed earlier gives us a fairly firm foundation for my statements.

Early 'Man' and the Geologic Record

One of the first arguments that will be presented against this theory is the lack for an accounting for the earlier forms of 'man' that predate the Biblical account of the creation of humans in Genesis chapter one. There have been fossils discovered of life forms that are similar in many respects to modern humans. Homo erectus is the form that walked upright as modern humans do based upon the study of the structure and probable function of these fossil remains. While there are other forms from differing time periods within the geologic record the differences between them and modern humans only grows greater as we search earlier in time. I have a possible explanation for the existence of these earlier life forms.

Creation: God's Laboratory

As we have discussed earlier the events of Creation as described within Genesis chapter one took a great deal of time to occur as God directed them. Our Creator experimented with many forms of plant and animal life. Sometimes, like in the mass extinction events that occurred during the Creation events, God decided to wipe the board clean of certain species while preserving others.

The only logical explanation of the above that can find both Biblical and scientific support is that God was experimenting with different forms. God would constantly improve and diversify His Creation by keeping some creatures while discarding others. While the forms that are closely related to man may have originated from an original pair and evolved through different forms in order to adapt in a progression, God eventually choose to wipe the slate clean of these forms some time before he created humans.

God may have even based Adam and Eve upon the designs of these creatures but we know that the first pair of humans were created in God's image and these other forms were not. If, as the Bible says in verses twenty-six and twenty-seven with verse twenty-six being a direct quote of God, we are created in God's image and after His likeness then we are modeled upon the perfection of God. God, prior to creating His first pair of humans, may have been experimenting with a wide variety of creatures to determine which of them would be the ideal forms that He would permit to share His earth with His created humans.

Science has recently refuted any connection between modern humans and the somewhat similar forms that existed prior to the creation of humans. DNA testing of the remains of these similar creatures shows no relationship to the DNA of modern humans. The further support for the lack of the connection between humans and these other forms is the elusive missing link. Science has never found a missing link, or transitional form, between humans and these other creatures because a relationship does not exist. In short, a missing link will never be found because it does not exist within the geologic record or anywhere else. Humans were a unique creation that may have had some characteristics that are shared by other creatures but not to the extent that one could even begin to assume that we were descended from them.

Humans were the final creation of all that God created during the vast time period of creation and have been here a very short time in comparison to the other creatures that we share the earth with. God wanted an ideal compliment to His human creation and took a very long time developing it. The creatures that were present at the time of the human creation were the products of natural selection, variation of species, and numerous adjustments made by God.

We have already discussed earlier the intentional role of creatures like the dinosaurs and why God chose to eliminate them. God, in doing all of this, was by no means being cruel or indifferent to any of His creations. Each and every one had a distinct and specific purpose. God also, as verses twenty-six and twenty-eight through thirty of Genesis chapter one states, wanted His human creation that was modeled after His image to have dominion, or to rule over, all the other creations of the earth.

Just like the example of oil previously mentioned in this book, God's creation has served humans since the first pair. Just one of the purposes of all that was created is the way that various plants and animals serve as sustenance for humans. These creations also serve to challenge our innate curiosity which was also modeled after God. It makes sense that God would have an innate sense of curiosity and the need to explore and experiment. It also gives a purpose as to why God created all that He did.

God spent a lot of time, as the vastness of time means very little to Him, creating, testing, experimenting, and observing His varied creations with the ultimate aim of creating a perfect environment for His human creation. Since the earth by its nature is always changing, God did create the perfect creatures by building-in to them the ability to adapt and change, or evolve, as time passed.

God and Error

God did not make mistakes during Creation as He is perfect in every way. He never created a creature that he did not intend to. Each and every one served, or still serves, a purpose within the framework of His creation. Science has found in the form of fossils some extremely unusual creatures. We even have some rather unusual creatures living today.

Take a look at Madagascar for an illustration of the creative solutions God has employed throughout the Creation events. There are thousands of unique creatures residing on this island just off the eastern coast of Africa. Many of these creatures are unique to Madagascar and are found no where else on the planet. Many of these creatures could not survive among the creatures found throughout the earth on other continents. The same can be said that if other creatures were introduced to Madagascar.

Many of the unique creatures indigenous to Madagascar would become prey to any new creatures. God may have broken this chunk of land from the African continent and used it as His laboratory. He may have also broken it off to separate and preserve some of the more delicate creatures for an as yet unrealized purpose.

Just as in the previously discussed discovery of oil and the way in which it has driven progress, God allows us to discover various items of benefit as we require them. We have to do research and create inventions on our own and as this progresses God gives us a piece of the puzzle now and then to aid us in our progress. God is the ultimate scientist taking great pride in observing His human creation as the steps of progress are taken one by one. It is beyond the scope of this book but many controversial moral dilemmas of today can be solved through an application of Biblical principals and modifying current thought to accommodate this process.

Theory of Use and Disuse Applied to Humans

There is another phenomenon that came to mind as I was working through these ideas of human evolution. I wondered about another distinctive characteristic of a part of the human population. I began to think about the narrowed eye openings of the Asian population as to where they may have originated or diverged from the rest of the human population. I believe that the answer to this phenomenon is rooted in the population of humans that are found further to the north in areas such as Siberia and the polar north. During most of the year these areas are found to have a covering of snow most of the year. It is also true that the further north a location is, the greater the angle of incidence of the sun's radiation.

Let us now look closely at the snow covering. We know that sunlight, when reflected off the snow, provides a dazzling array of sparkling light. When the sunlight is reflected off the snow into our eyes we find ourselves squinting our eyelids to reduce the amount of light we allow to enter our eyes. This idea is similar to looking at the effect of staring directly at the sun. It becomes painful very quickly and we turn away or close our eyelids to protect them. We also know that if we allowed our stare to continue we would damage our eyes to the extent that we could experience temporary and even permanent blindness.

Another support of this idea of the reflection of the sun's radiation off the snow can be supported by observing snow skiers. Usually the only exposed skin on a properly equipped snow skier is the face. Sunburn is a frequent occurrence among skiers due precisely to the reflection of the sun's radiation off the surface of the snow.

Back in the snow covered north we have humans that have to squint constantly to be able to function in this environment of reflected sunlight. If we apply the Theory of Use and Disuse here we can say that over time and generations that the squinting of the eyes will have some sort of effect on the structure of the eyelids. The less that the eyelids are opened to their fullest extent the smaller the measurement becomes of a full extent opening.

Now, very gradually and many generations later, we find people with a narrow eye opening when compared to the rest of the human population. This characteristic is found among the Eskimo people, the people of Siberia, populations of Tibet and Nepal, and the Asians indicating a direct ancestral relationship among them. This also would support the idea that the origins of this characteristic are farther north than the Asian nations of China, Japan, and the far east indicating that there was a vast migration of part of the northern population to the south. This migration may have been instigated by a major climactic change such as would have been brought about with the advent of the Ice Age.

Science Makes Connections

I am attempting to permit you to see the many connections that science is capable of supplying us with. We will not go into a discussion of how the Asian migration to the south would be a support of the idea of the Ice Age. I am only illustrating the fact that science works in this way. Science makes assumptions and sets out to uncover support or refute for those assumptions. Science is also never ending and can be compared to the branching of a tree. Look to what we just did above. We moved from the narrowing of eye openings to the Ice Age in relatively short order.

Scientific and Biblical Tests

Since we examined the phenomenon of narrowed eye openings in a scientific manner we can say that we have a scientific basis for the original idea. Again if we apply the

Biblical test we really find no mention of this phenomenon. The next test is to see if our idea refutes what is found within the Bible and we find that it does not. We now have a scientifically and Biblically sound hypothesis concerning the evolution of narrowed eye openings in a portion of the population.

While I can already hear the collective sigh from those wondering how I could come to such preposterous conclusions I would ask you but one question: What are your theories on the subject of human evolution? What I have proposed answers many questions that have been pondered throughout the ages. Just like any other theory mine are also subject to additions, modifications, and deletions as new and better ones are brought to light. It is my sincere hope that science finds stronger support for these theories and that, if they are erroneous, they are proven to be so. This only opens the stage to the new and improved ideas that come from progress.

What I have proposed is what I believe to be the best explanation based upon human history, current scientific methods, and the Biblical account of Creation. Since my conclusions are based upon such principals I believe that they would be acceptable to the vast majority. They are, like any other theory, subject to debate. Debate that leads to a better and more complete theory is productive debate. Debate that polarizes the sides against one another, such as the current Evolution/Creation debate, is counter-productive to the progress that benefits humankind. We see direct evidence of this by observing the numerous battles from within the local school board all the way up to the United States Supreme Court and everywhere in between.

Apes, Chimps, Monkeys, and Man

Many believe that humans have a close connection to the closely related apes, chimpanzee, and monkeys. It may be because these creatures sometimes walk on their hind legs or that some of them have been observed to modify an object for use as a tool or display other human-like behavior. While these creatures may be among what could be classed as some of the higher thinkers of the animal kingdom they have been shown, through examination of their DNA, to have no direct relation to modern humans.

Some will argue that there is a small difference between a chimpanzee's DNA and a human's DNA. Although I will concede that that may be true it is these differences themselves, no matter how small, that distinguish them as unrelated. To allow that to compel one to believe that humans were descended from chimps is preposterous. Just go back to the fact that there is no intermediate, or transition form, between humans and chimps that has ever been found.

When you look closely at these ideas what one finds is further support for the idea that humans and the chimps are in no way related. There may very well be a relationship between the apes, monkeys, and chimps due to the fact that they are all mammals. Yes, humans are also mammals but again all you have to base a direct relationship upon are suppositions as there are no transition forms which would be the irrefutable support of this idea. I would further turn that idea upside down by stating that the lack of any transition forms is irrefutable support that no descent relationship exists at all.

Chapter VII

The Theory that is Not a Theory

The Theory of the Origin of Species

The Theory of the Origin of Species is the main point of contention between Creationists and Evolutionists. The main controversy surrounding this theory is that it attempts to trace all species back to one common ancestor and directly challenges the Biblical account of Creation. What occurred in the formulation of this theory is that the Biblically and scientifically supported theories of natural selection and variation of species were mixed with unfounded and unsupported assumptions. Remember, as I pointed out earlier, that most of the unfounded and unsupported assumptions come not from Darwin. The 'man from apes' and 'primordial soup or slime' ideas come from the last several decades.

Perhaps the most important point to remember here is that this part of the theory, as we know it today, bears little resemblance to Darwin's original theory. Most of the terminology, ideas, and assumptions we use today when discussing the origin of species did not exist at the time and were not included in Darwin's original ideas. Science, in combination with sensationalism through the media which is demanded by the public, has translated the original idea into a very complex collection of assumptions.

The key evidence to support this part of the theory has never been found because, in fact, it does not exist. The Evolutionist asserts that the original creature from which all life forms evolved can be proved by observing all the relationships among all the species that we have today and tracing those back until they eventually merge to the single creature. There are even mixed views on how this is supposed to have happened.

Two Sequences or Forward and Reverse

In one camp you have a group that believes all life arose from the water. This idea would start with a water bound creature morphing to amphibians then transforming to land animals. The illustration of this would be that fish with the legs and feet that has become a somewhat popular symbol on the trunks and bumpers of cars. I propose that the vast majority of those displaying such symbols have absolutely no idea as to what the symbols represent: the elusive transition form that has never been found.

The other camp consists of those that think that land animals became amphibians and then transformed into fish. If either one of these assumptions were true then the ultimate supportive evidence of them would be the fossils of the intermediate or transition forms between them that have never been found.

The intermediate forms, if they existed, like the others that have been present on the earth, would have left some trace of their existence behind. The problem with these assumptions is that these intermediate or transition fossils or remains have never been found. The fossil that I would like to see personally is the little fish with the legs and feet that was discussed earlier.

The common argument for the lack of these transition fossil forms is that the geologic record is incomplete. We have unearthed remains and fossils, both huge and miniscule, that are older, newer, and within the same time periods that these transition forms were purported to have existed.

Another idea is that these intermediate forms were too small and their remains would not have survived long enough to make their imprint within the fossil record. Of all the creatures that have left evidence of their existence within the geologic record not a single example of a transition form is to be found. Where is the evidence, in the form of transition creatures, which would support a common ancestor?

Darwin and Creation

Charles Darwin stated that he did not believe that the variety of species we observe on the earth today was a result of divine creation. He instead defined God's creation as the result of natural occurrences. So shocking were his beliefs to even himself he only attributes a line or two of his book to his unfounded assumption. Even Darwin himself cannot explain why a transition form's fossil is so elusive choosing to affix blame to an incomplete geologic record.

It is true that God created a finite number of creatures during the creation event and that natural selection, variation of species, climactic change, and countless other factors caused the diversity we see today. These natural events themselves that affected the original creations and created the variety of today were designed and put into motion by God also. His purpose: to create a perpetual creation that would grow, adapt, and survive forever.

It is the nature of science to have some kind of reproducible experiment or often observed piece of evidence to support an idea before it can be called a theory. While parts of Darwin's theory are both Biblically and scientifically sound, his origin of species assumption of a common ancestor does not fit the standards of either.

What science and the Bible do support is that there were a number of species that appear in the fossil record at approximately the same time and that over the course of time different creatures have existed at different times. These variations are the product of natural occurrences such as reproduction, climatic changes, natural selection, and variation of species.

Our only conclusion here that has any merit is that Darwin took a few valid ideas, tacked on a suspect and unsupported assumption, and called the collection as a whole a theory. It also must be recognized that Darwin himself thought that including a directly stated theory of human evolution was too controversial to include in his work.

Much of what people in general believe to be true about human evolution came not directly from Darwin's ideas but directly through different interpretations and misconceptions passed down throughout the years. To satisfy both sides all we have to do is examine the individual ideas of the theory, apply the scientific method to them, and discard whatever cannot be supported. To be sure that what we do not discard is acceptable we then apply a Biblical test to see if those ideas can exist concurrently within the framework of God's Word.

In a nutshell, this part of the theory does not stand up to scientific scrutiny. This part of the theory also disagrees with and directly refutes what is found within the Bible. Since it fails both tests this idea must be discarded and attempts to find an alternative explanation must be pursued by science.

Chapter VIII

Time and the Bible

Creation: Time

If we are to take the account of creation as presented in the Bible literally, then we must believe that on the third day, according to Genesis one verse nine, God caused all the water upon the earth to be gathered into a single place revealing dry land. This verse, if taken literally, accounts for one ocean and one land mass or continent.

Science supports Genesis one verse nine exactly as it is written with the single continent/single ocean and continental drift theories. The primary issue between creation and science is now narrowed down to only be one of time. One continent could not have broken into seven pieces and drifted to their current positions in the time allotted for within the Bible if the time component of creation is to be taken literally. Based upon measurements of the current movement rates it would take at least 250 million years for the single continent to break into pieces and these pieces end up where they are presently.

The next time issue presents itself in verses twenty through twenty-two. It is the fifth day and God is creating the marine animals and the fowl. Both of them God blesses and directs to reproduce to fill the water and the earth. Now if we make the assumption that God created two of each kind, male and female, as he did humans, then there is a time component associated with reproduction's processes.

Verse twenty-one of Genesis chapter one mentions that the marine animals:

> 21 "...which the waters brought forth abundantly, after *their* kind..." (Author's italics).

Abundantly, after *their* kind and the waters brought forth statements both indicate that the marine animals reproduced in the waters to fill the waters of the earth themselves. The process of reproduction varies greatly among the marine animals and to reach a point of abundance in the waters of the earth would require a great number of generations of each animal species. Again if these verses are to be taken literally then the only discrepancy between science and the Bible is time.

There exists, in Genesis chapter two verse four, an explanation of these time discrepancies as provided by God Himself following the description of the seventh day of creation:

> 4 "These *are* the generations of the heavens and of the earth when they were created, in the day

> that the LORD God made the earth and the heavens…"

To find this description of "These *are* the generations…" immediately following the account of the seventh day implies that the time component is not to be taken literally. The term generations implies reproduction and reproduction, in the case of generations, takes a vast amount of time. Furthermore, the generations required to fill the waters, sky, and earth would be so great a number that it would be difficult to comprehend.

You need to note that not only would each species' population would have to grow exponentially with each generation, they would also be faced with the challenges brought by the controls of natural selection and variation of species. The account of the creation of humans, in much greater detail than is conveyed in the first chapter of Genesis, follows this time description in verse seven of Genesis chapter two. God was attempting to clarify two key points from the creation event that Genesis chapter one describes.

After *Their Kind*

Creation was an event that was designed and conducted by God but certain events were initiated and the course of natural reproduction, selection, and variation of species was allowed to follow their courses. There are two references to evolution in verses twenty-one and twenty-five of the first chapter of Genesis:

> 21 "And God created great whales, and every living creature that moveth, *which the waters brought forth abundantly, after their kind...*"
>
> 25 "...and cattle *after their kind...*" (Author's italics and underscore for emphasis).

"After their kind" directly infers reproduction and that each creature's eventual existence in their respective forms were not the direct result of His creation but rather a product of the species, the environment, selection, and variation that He created and directed over a period of time. This is further reinforced by the statement: "the waters brought forth abundantly" which indicates that creation was taking place through reproduction on its own.

You can now deduce that if the waters brought forth abundantly, as God said they did, then a great deal of time would have already passed. Remember that reproduction is occurring with the controls of natural selection and variation working against it and the process went on until an abundance of these creatures existed *during* the events of Creation. God intended for nature to take its course as that was a part of His intentional and intelligent design.

God repeats these same instructions to Noah in Genesis chapter six verse twenty:

> "Of fowls *after their kind*, and of cattle *after their kind...*"

Again the 'after their kind' statement to emphasize the fact that these creatures were not directly created by God but were products of His original creations that had changed in some ways through reproduction, natural selection, variation of species, and time.

Several thousand years had passed since the original creation as evidenced by the genealogy given in Genesis chapter five that traces the generations of Adam through Noah. God again is conveying the passage of time to the reader since now it can be defined easily using the names and life spans, measured in years, of the descendants of Adam and Eve which were His original creations. Prior to the creation of man there was no easily definable method to convey the passage of time except Genesis chapter two verse four where God speaks of "...the generations of the heavens and of the earth when they were created..."

Science Supports the Biblical Account of Creation: Sequence and Time

We now have scientific support for some of the key events that are included in God's account of creation. Science supports the existence of one continent and one ocean as described in Genesis one verse nine. Science also supports the idea of change and diversity of species over time as described in the 'their kind' statements we discussed previously. If one accepts the idea that, although not defined precisely, there were passages of varying lengths of time between and during the creation events as both God's description and science supports, then a joining of evolution and creation can be considered.

Intentional Design: Controls

God had the foresight to know that if He did not place some sort of controls on His creation that it would destroy itself through uncontrolled reproduction and the resulting depletion of resources. He also knew that the earth, as He designed it, would have seasons and the accompanying changes in climate. He further saw the need for his creations to be able to change – or evolve – to better adapt to changing conditions as time passed. These built-in controls would permit His creation to survive perpetually.

Lesser Light/Moonlight/Reflected Light

Let us now look carefully at the "...lesser light to rule the night..." passage from Genesis chapter one verse sixteen. God is clearly referring to the moon since the stars are mentioned separately. The moon produces absolutely no light of its own; it only reflects the light of the sun. If we take the passage literally we have to conclude that the moon generates its own light. We know that this is not true and we have further support of the idea that the account of Creation in Genesis is delivered, at least partially, in parable.

We only find more support for the idea that time was a factor not precisely defined in the account of Creation. God was teaching, partly in parable, when He spoke of the 'days' of Creation. One need look only to Genesis chapter two verse four to find God's explanation:

> 4 "These *are* the generations of the heavens and of the earth when they were created, in the day that the LORD God made the earth and the heavens,..." (Italics in original).

There were no humans present during the Creation events to record them. We have only the divinely inspired words of God to explain the Creation in terms that the humans of the time could comprehend. Time, as we define it, is divided into minutes, hours, days, years, and so on. The way we define time is of human origin and therefore time was defined in terms that could easily be understood by the people of the time in which the Bible was written. God also, in His Divine wisdom, wanted to convey His concept of the Sabbath day.

In reality the events of Creation did take time, and a great deal of it. God was providing us the sequence of the events of Creation – not a literal time period. If each of those time periods that He names symbolically as days contain millions of years apiece then our Sabbath would be longer than human history itself. God spelled out the idea of the seven day week and that He intended the seventh day to be the Sabbath.

We now have a great deal of evidence compiled supporting the idea that the Creation event occurred over a period of time that is not defined. This idea is supported by numerous pieces of evidence found within God's Word and supported by science with the inverse also holding true.

I am not, and never have, disputed a single word written in the Bible. I am disputing a long held interpretation of Genesis chapter one. To summarize, I am proposing that since numerous examples of the use of parable are found in the Bible that Genesis chapter one also follows the parable teaching model. The account of the Creation, as far as time is concerned, is taught in parable. God was defining the sequence of events and impressing the importance of the Sabbath with His parable.

Chapter IX

New Biblical Theories Supported by Science and the Bible

We now arrive at a juncture where it is necessary to explore some new ideas, or hypotheses, to explain some of the conclusions I have made. The following ideas took many years of thought, study, and research to get to the point where I could tie everything together in a logical way utilizing scientific methods. I also, before sharing them with you, had to ensure that they were Biblically correct, sound, and did not, in any way, go against my faith. I believe I have succeeded in that endeavor, and hope that you agree.

The Theory of Twos

God created humans by creating one male and one female with the intention that they 'be fruitful and multiply' to fill the earth. The account of Noah and the great flood that destroyed all life on earth finds God instructing Noah to collect all species in twos, a male and a female of each, to replenish the earth after the flood.

In each of the above cases it was only necessary, according to God's plan, to have a male and female of each species to populate the earth. He also, according to Genesis chapter one verse twenty-two and verse twenty-eight, blessed His creations and instructed them to be 'fruitful and multiply' to replenish the earth and fill the waters.

I propose that God designed and implemented the idea of creating two of each species, male and female, to carry out His direct process of Creation. The process of replenishment would occur through reproduction, natural selection, and variation of species just as He designed and directed it to. The precise number of species that God created during the period of Creation is not given and cannot be determined. God does not provide us with the number and humans did not exist during the creation of all the original animal species to count them.

Theoretically speaking, it is possible to produce a population capable of world wide distribution that originates from a single breeding pair. This process, however, would encompass a vast amount of time. In addition, the effects of natural selection, variation of species, and other natural occurrences would hinder any extraordinary explosive growth. There is nothing in the Bible to refute the Theory of Twos and it finds support on a scientific level. Now you have an important piece of this puzzle we are trying to put together: the Theory of Twos.

The Theory of Angels and Time

While writing this work I found another apparent mystery with verse twenty-six in Genesis Chapter one:

> 26 "And God said, Let us make man in our image, after our likeness..."

God is using the term 'our' in referring to man. "Our image" and "our likeness" indicates that God is clearly speaking of Himself and another being or beings that are like Him and present with Him as He was creating man. I also base this on the fact that Satan is present shortly after creation in the Garden of Eden and, being one of the fallen angels, a number of angels had to be present before or during the creation events.

The idea behind this theory asks that you agree that the first two verses of Genesis chapter one to have no defined measures of time. If you also can agree, based upon the previous ideas, that the time component of the Creation event is not precisely defined, then you have time for Lucifer to challenge God, as found in Isaiah Chapter fourteen verses twelve through fifteen.

Within these verses we have the account of Lucifer challenging God and being cast down. Having no time constraints within the Creation event allows Lucifer time to challenge God, be cast down, and then be found shortly after the appearance of man in the Garden of Eden. Again we find more support of the idea that time is not defined during the Creation event.

The Biblical 'his' Theory: Political Incorrectness

Another issue that I have given much thought and prayer to is one that I have toiled with for a long time: the use of the word 'his' throughout the first Chapter of Genesis. At the time the Bible was written, and especially so when it was translated, the age of politically correct terminology was still far in the future and the male was considered to be the dominant of the human species. Almost universally, when speaking of an individual without naming them, the words used were he, him, and his. Look at Genesis chapter one verse twenty-one at the end of the verse:

> 21 "...and every winged fowl after his kind: and God saw that *it was* good."
> (Italics in original text)

And also look at verse twenty-four:

> 24 "And God said, Let the earth bring forth the living creature after his kind, cattle, and creeping thing, and beast of the earth after his kind: and it was so." (Author underlined direct quote)

Now to tie this idea together look at part of verse twenty-six:

> 26 "And God said, Let us make man in *our* image, after *our likeness…*"
> (Author's italics for emphasis)

You need to especially note that verse twenty-four starts with "And God said" indicating that He is speaking – a direct quote. To "Let the earth bring forth" indicates a process God is directing to take place but not actually doing it Himself, hence, reproduction. Also contained in God's direct quote twice, and referring to "the living creature" and the "beast of the earth", we find "after his kind". To whom is God referring to when He says "his kind"?

God is speaking of these species, in particular the male of these species. In today's time of political correctness the term would probably have been 'after *their* kind'. We know that if God is creating humans after His image and likeness then He cannot be also as a bird, fish, and beast of the earth. Nevertheless God in speaking, defined by His direct quotes, would not use 'his' when referring to Himself, He would have used 'My' or 'our' as He does in verse twenty-six.

Chapter X

Creation and Evolution Joined

Creation Joins Evolution: Sequence

According to the first chapter of Genesis the first day ends with a formless earth, a heaven, water, and light. God further defines the light as day and the darkness as night before the end of the first day. God spends the second day creating the firmament (American Heritage Dictionary: "The vault or expanse of the heavens; the sky.") to divide the waters of the earth from those of the heavens.

Up to this point we can take the account of creation on faith and science prior to Genesis chapter one verse nine because we have, to this day, an earth, water, and a sky. The third day, starting with Genesis chapter one verse nine, gives us an earth with one continent and one ocean. This verse is entirely supported by scientific evidence and the Theory of Continental Drift. Also on the third day were formed grass, herbs, and fruit trees with seeds in the fruits which also can be taken literally supported by the fact that they all exist to this day.

The fourth day is when God creates the Sun, Moon, and stars. He describes them by designating the:

"...greater light to rule the day, and the lesser light to rule the night..."

There is, for the first time, a bit of a mystery to be found here. In Genesis one verses three through five God creates light, divides the light from the dark, and names the light "Day" and the dark "Night". It is beyond the scope of our discussion but the source of this light before the sun, moon, and stars were created is not explained.

Verse twenty begins the fifth day in which God creates the marine animals and directs them to multiply to fill the waters. God also creates the fowl and likewise directs them to multiply over the earth. The last working day – the sixth – of creation begins with God creating the "beasts" of the earth or simply land animals. He specifically mentions cattle by name providing for the first specific appearance of land mammals. It is also on the sixth day that God creates humans, one male and a female and directs them to multiply to fill the earth. Insofar as the sequence of events there is agreement, for the most part, between creation and what science supports.

Dinosaurs, Fossils, and Geologic Time

Science supports the existence of the dinosaurs through the direct evidence of the fossils and skeletons that have been found. While the Bible does not mention dinosaurs directly their place is found in verses twenty-four and twenty-five of the first chapter of Genesis:

24 "And God said, Let the earth bring forth the living creature after his kind, cattle, and creeping thing, and beast of the earth after his kind: and it was so."

25 "And God made the beast of the earth after his kind, and cattle after their kind, and every thing that creepeth upon the earth after his kind: and God saw that it was good."

Again, as we have previously discussed, the 'their kind' statement implies reproduction, natural selection, and modification or variation from the original creation. The important concept to grasp here is that these processes all require time. If one allows for this time component and recognizes that the 'their kind' statement in fact describes a modification from the original then we can join creation with what science supports easily.

Whether God directly created the dinosaurs or they were a product of modifications to His original creations is not spelled out. Science would support the theory of larger creatures modifying to smaller creatures rather than the inverse as being the most likely sequence. Based upon this, one could deduct that the 'beast of the earth' references in verses twenty-four and twenty-five may refer directly to the largest of the dinosaurs.

God designed and created the original groups of species and a set of controls that we call variation of species and natural selection exactly so there would be a diversity of species and, more importantly, to place a control on exponential population growth. Without natural selection population would grow unchecked until all the resources were consumed. If all resources were consumed mass extinction would be the end result, to the extent that no life would be sustainable upon the earth at all. Natural selection and variation of species can only be the result of a Divine and intelligent design that has been present and in operation since God created the first creatures to inhabit the earth.

We will return to the specific examples given in Genesis chapter one verses twenty through twenty-two and verse twenty-five. Verse twenty has God directing "...the waters bring forth abundantly the moving creature that hath life..." indicating that the water – not God Himself – has the duty to "bring forth abundantly".

The above indicates that the process that was initiated by God, the Creation event, was to carry on the process, through reproduction, natural selection, and variation of species to "abundance". This view is clarified and restated more forcefully in verse twenty-one.

> 21 "And God created great whales, and every living creature that moveth, which the waters brought forth abundantly, after their kind..." After God creates the great whales and the other marine animals the "...waters brought forth

> abundantly..." The pairs of species that God created were directed to reproduce to abundance.

The key to variance occurs in the next three words of verse twenty-one: "...after their kind..." This *after their kind* statement, directly states that the abundance of creatures brought forth within the water was the product of the creatures themselves and not a result of God's direct act of Creation.

Look at verse twenty-two as it only reinforces what we have deduced from the previous two verses when God blesses his creations and directs them to

> 22 "...be fruitful, and multiply, and fill the waters..."

God created the originals but directed them to "...fill the waters..." God's grand design required only two of each created species to begin the process which He directed to occur. He repeated these same instructions to the first pair of humans, male and female, in verse twenty-eight.

We also have the Theory of Human Evolution discussed in chapter VI to add here. The theory is Biblically supported and it supports what is found in the scripture. Further the theory is based upon scientific principals and methods and therefore meets the tests from both sides.

Throughout our study of these issues we have constantly applied two tests to each and every idea: Does the Bible and science both support it? We have found that through applying both tests to each item we have only one item out of the Theory of Evolution that could not be supported by the Bible nor science: the idea that all life arose from a common ancestor. The other ideas contained within the theory were shown to be both Biblically and scientifically sound. This leaves us with a dilemma: What do we do with the theory as a whole?

Renamed Theory

Since we have defined and tested all of the ideas both Biblically and scientifically, there is only one name that would describe our theory properly:

The Scientific Theory of Creation

There is a slight problem with this name due primarily to issues of separation of Church and State. Since the primary purpose of this book is to bring two diametrically opposed sides of a debate together, certain neutral terms must be used to satisfy the masses. We now must rename the theory:

The Scientific Theory of Intentional & Intelligent Design

No, I did not take God out of the theory. I have never, in any way, disputed any word within the Bible. It is true that I challenge long held ideas about parables and literalness. I also have used the standard and current scientific methods in my analysis of every idea presented here.

What I have accomplished, when these ideas are carefully examined, is to remain true to both sides of this debate proving that a joining of the sides is possible. It has been the most perplexing task to keep this discussion as unbiased as possible. I believe many people have shared these thoughts to some extent. Perhaps the difficulty I face in creating this work is the very reason why no one has taken it on in this manner.

Modern society has required that a division be placed between Church and State. A recent battle in my hometown that illustrates this point perfectly concerns our City of Edmond Seal. Divided into four fields one of them contained the Cross symbol. One person, a non-believer, initiated a lawsuit that resulted in a plain white sticker being placed over the cross. The amazing support shown for the symbol was illustrated by the act of most of Edmond's residents placing small white crosses upon their lawns all over the city. Not that it is important but the person who initiated the lawsuit now resides in another state. This illustrates the level of division between the sides and also shows the capability of a single person to effect a change that impacts all of society.

We have traveled a great distance in the over eighty years since the Scope's trial when the courts fined teachers for presenting ideologies that refuted or did not include Biblical principals. Now our courts will find fault in teachers who present any idea with a connection to the Bible within their classrooms. There has been a complete turn in the reverse from where we were not too long ago.

These issues are what provide me with the greatest challenge I face as I attempt to deliver a theory that will not be struck down just because of separation of Church and State issues and not be discarded by believers either. When reading the theory the first thing one will note is that it allows for universal applicability. Christians will see the hand of God at work within it, non-believers will be able to leave that open, and Buddhists will be able to put Buddha in it, and so on. It is a truly universal theory that can be modified by each individual to their own system of beliefs.

My fellow Christians will take issue with me on the above but I will refer them to my main purpose in writing this book: to end the debate and bring the sides together. Christians can see that I did not leave God out of it because I have repeatedly referred to the Bible and shown mutual support between my theories, the Bible, and science. Read the theory carefully and you will see that there is a place for God among every point that is made.

I sincerely hope that due to the universal language that the theory is presented in it will gain serious consideration from all no matter their faith, or lack thereof. It may even serve to bring others to consider God and eventually find their way to acceptance.

There is no other way to present this theory and hope for a universal acceptance if it is biased one way or the other in anyone's eyes. The fact that it is based on scientific principals should make it acceptable to all. The fact that it also is based upon Biblical principals should make it acceptable to Christians. These are the purposes for my attempt to create a truly universal theory acceptable to all.

Chapter XI

The Scientific Theory of Intentional & Intelligent Design

The Scientific Theory of Intentional & Intelligent Design contains a modification of the Theory of Evolution based upon both scientific and Biblical support and evidence. The theory also includes theories of others that were included in the Theory of Evolution as well as new theories that were developed by me. I incorporated all the theories into one as they all pertain to the subject at hand: the joining of Creation and Evolution. The Scientific Theory of Intentional & Intelligent Design contains the following:

1. The idea that there was once nothing and then something – or a beginning – as stated in the Big Bang Theory is Biblically and scientifically sound. The theory proposes that the universe started as a single point that expanded and is still expanding. Contained within is the entirety of the universe as we know it.

2. The idea that the great diversity of life that exists today could NOT have come from a common ancestor is Biblically and scientifically sound. There exists no valid support of the idea that all life can be traced to a common ancestor.

3. The idea that the Theory of Natural Selection is Biblically and scientifically sound. This idea is best summarized as 'survival of the fittest' which is an intentionally built-in control against explosive population growth.

4. The idea that the Theory of Variation of Species is Biblically and scientifically sound. Variation of species is an intentionally built-in mechanism that allows each species to adapt to the ever-changing conditions present on the earth.

5. The idea that the Theory of Use and Disuse is Biblically and scientifically sound. This theory is closely related to variation of species as it is also an intentionally built-in mechanism that allows each species to adapt to the ever-changing conditions present on the earth.

6. The idea that the Theory of Twos is Biblically and scientifically sound. This idea states that only a single male and female of any species is needed to start the process of world wide distribution of a species' population.

7. The idea that the Theory of One Ocean/One Continent is Biblically and scientifically sound. Early in the earth's history the earth consisted of one large land mass and a single ocean surrounding it.

8. The idea that the Theory of Continental Drift is Biblically and scientifically sound. The earth once consisted of a single land mass that broke apart with the pieces drifting over a vast amount of time to the current locations of the continents.

9. The idea that the Theory of Intelligently Designed Human Evolution is Biblically and scientifically sound. Since the first pair of humans were the only humans created and the human population consists of many races, the different races are descended from the original first pair. The differences between these races of the human species are the result of environmental conditions present at the locations where they lived. The environmental factor with the greatest impact upon differences between the races is the level of exposure to the sun's radiation.

10. The idea that the events of creation occurred over a period of undefined time is Biblically and scientifically sound. The Creation event described within the Bible is accurate as to events and sequence. The term 'day' is used only as an illustration and does not literally define the passage of time during the events of creation. Support for the idea of time being defined in a parable comes from the many instances within the Bible where parables are used to convey difficult to understand or grasp ideas and concepts.

Once general agreement has been reached through numerous attempts to support or refute these ideas, both sides of the issue can now have a common ground from which to build a new relationship between science and the Bible. There now exists a scientifically and Biblically supported theory that will not find opposition within the home, church, or the classroom from any side once the tests of scientific and Biblical support are carefully applied and examined. In other words it is a universally understandable and universally acceptable theory.

Why this theory is important to science

Science has an important place within our society. The institution of science finds itself at odds with an ever-growing proportion of the population as time passes and science progresses. These differences result in fierce debates on issues such as evolution, stem cell research, cloning, and other issues that are mostly ethical in nature. At the same time, we find that a joining of forces is critical to our success if we are to overcome this ever present stumbling block to further advancement for the benefit of all. The origins of the ethical beliefs within society are rooted within Biblical principals that have withstood the tests of time and change.

The people demand that science follow these same ethical guidelines that are followed by them. We must first always find an ethically sound basis for our new technology. If one is not apparent then we must search further or make modifications to enable the ability to pass such ethical tests. When science applies these ethical principals to their theories, before any presentation to society, there will be found a vast support and acceptance of ideas old and new. This universal support will serve as a platform for further, and far more aggressive, excursions into the realm of the unknown for the benefit of all.

The more apparent benefit lies in the fact that we now have a mutual platform from which to educate our future generations of scientists. These new scientists will have the tools and methodology from which to maintain an ethical and scientific search that will find universal acceptance. The only way to accomplish these goals is for the scientist to ensure a correct translation of terms into ones that are easily understandable to the masses.

If the people can not understand your ideas and theories they will not accept them. You further leave your ideas open to interpretation that almost always results in misconceptions and unfounded connections to more sinister intent. This will be either instigated or perpetuated by the media and only because you neglected to define your terms in an understandable format.

Translation of terms into universally understandable terms is far from easy as I have found through writing this book. The key to correctly translating terms is to present them to the lay person. If they do not understand and are full of questions then you have not accomplished a meaningful translation. Unanswered questions lead to translation by those not qualified to do so and this is where controversy through misconception is born.

Why this theory is important to Christians

We have come to a point of such severe division between science and religion that we are impeding progress that is necessary to the survival of God's creation. If we only demand instant integration of Biblical principals into the whole of science without providing a means to do so then we will be at odds with science on a constant basis.

We, as Christians, must allow for a method by which science is capable of applying our required tests to their ideas. This joining will require a deeper investigation and understanding of our world utilizing both scientific and Biblical tests along the way.

Perhaps the greatest importance of this theory to Christians is that we now have a Biblically and scientifically sound theory to teach our children. The best feature is that the theory is slightly adjustable in terms to deny the separation of Church and State advocate a means to keep it from our children's educational institutions.

Translation to terms appropriate to the Church and secular educational institutions is accomplished simply by inserting God into the appropriate places. Simply stated this theory is universally understandable and universally acceptable to the masses whether they be believers or non-believers. We must not exclude any particular group from this news and it does have the potential to draw others closer to the truth.

Why this theory is important to non-believers

This theory with its neutral terms allows the dissemination of a valid theory without offending any one particular group. It also allows for a doctrine free education for your children. They will, at some point in time as they mature, come to their own conclusions as to if they want to attribute the events of creation to a particular entity or not. Some may wish to agree that this is how it all happened and leave the identity of the intentional and intelligent designer to remain a mystery. However you choose to look at it the theory can be accepted without a compromise of your particular beliefs.

Why this theory is important to society as a whole

The majority of earth's human inhabitants believe in some form of a higher power. In the United States the vast majority of believers subscribe to God as their higher power. The division between the scientific community and the majority is only growing and the end result is unimaginable for both sides.

Neither side has to compromise their inherent beliefs in order to agree on this issue. We must, for society's sake, find a solution that can satisfy both sides and eventually bring them together. The division, if permitted to grow, will result in a gradual loss of support that will have devastating effects upon all of society. We will progress toward a unified effort if we continue to produce and demand universal understanding in the name of universal acceptance.

Conclusion

This book was written with the intent that both sides find a common ground in the debate between Evolution and Creation. The greater challenge is to ensure that no part of the population be excluded based upon their beliefs. True believers have discounted many ideas purported to be based upon science because they directly refute what they believe in.

As you have just learned, because of one unsubstantiated claim and the many that grew from it over time, an entire body of valid ideas has been essentially discarded due to a misconception. Was the misconception cause by the scientist himself or by the perpetuated misconceptions that grew more numerous over the years? As Christians, it is our duty to reject that which refutes the Word of God. In discarding these ideas out of hand, we have missed the opportunity to incorporate many valid ideas into our thinking and our lives that could be of great benefit.

As Christians we must ensure that what we do discard is carefully examined according to a Biblical standard. We must attempt to find direct correlation or rejection based upon what is written in the Bible – not how what is written is interpreted. The key to interpretation is to ensure its correctness based upon the tools that God has provided.

God provides one of those tools in the form of science itself because it has provided so many benefits to all of God's creation. When applying a particular interpretation to any passage within the Bible, test to see that it stands up to scientific scrutiny. Once this is accomplished an interpretation can be made that has support from science and would, therefore, be more acceptable to more people. Is that not what God would have intended? A spread of His truth to all the people of earth seems to be God's intent. This can be more readily accomplished if we also apply the method of universal understanding to bring about a universal acceptance.

Both sides have failed, until now, to examine the theory as I have with the specific intent to bring the sides together in agreement. As scientists we must strive to observe and investigate applying all tests and experiments conceivable to support our claims. If we are to hope for a universal acceptance of all science has to offer then we must constantly apply not only scientific tests to our ideas, but the Biblical tests as well.

We will find that although our work is doubled, we will find the universal acceptance and support that has been lacking for our ideas amongst the vast majority of the populace when we apply both tests to our conclusions. For way too long science has found no place for religion and, in return, religion has gradually began to discount science.

The scientific community will find, through the incorporation of Biblical testing, a far greater and receptive audience. The religious community, which has been growing drastically in recent times, will demand such measures be taken. It is no more complex than applying a code of ethics, morals, and values to all of our research.

If, at any time we find our conclusions at odds with widely accepted values and beliefs, we need to search further. A supportable answer that can satisfy everyone is not that difficult to obtain. I have applied such techniques within this book, and although it took many years of study and research to arrive at these conclusions, I have supported the idea that it can be done.

Although it is beyond the scope of this book, I feel it necessary to address the issues that created this debate in the first place. The reasons for the division between science and religion are numerous. Perhaps the most important reason lies within the failure of science to convey ideas in terms that can be understood.

A vast portion of the population has been left with a growing body of technical jargon that easily leads to misconceptions which, in turn, inevitably leads to debate. While debate is a necessary and normal occurrence, the type of debate I refer to is that debate which is created solely due to a lack of understanding. Education of the masses in understandable terms must be made a priority if the chasm is to be filled. I have illustrated exactly that within this book.

Many people believe that the sole premise of the Theory of Evolution is that humans evolved from monkeys. A serious lack of understanding creates this line of thinking. The media makes no attempt to translate the jargon of science as they tend to skim for the controversial points to generate attention grabbing headlines. The media cannot be held at fault for this as they are only catering to demand. It is also not the responsibility of media to translate for science.

Science is to be held accountable for these inaccurate translations as it should be a function of science to deliver its findings in a manner that is thoroughly understandable to the masses. We have, for far too long, been concerned only of our colleagues' reception and interpretation of our ideas leaving the rest of society to come to its own conclusions. The sole result of this lack of understanding is misconception, debate, and rejection of our ideas by society.

Universal Understanding Brings Universal Acceptance

Universal acceptance should be the goal of all scientists. The only way to gain universal acceptance of our ideas is to first ensure a universal understanding of our ideas, concepts, and theories. If your idea or theory excites you, try explaining it in understandable terms to someone not of the scientific community. When you can generate a level of excitement and interest comparable to your own in that person, you have found a way to deliver universal understanding and will automatically gain the universal acceptance and support that is so desperately sought.

I believe that I have also presented plausible and valid Biblically and scientifically supported theories. These are theories that are backed with explanations that are universally understandable to the vast majority of the population. With this universal understanding can come a universal acceptance which can finally put to rest a debate that has invaded our society for far too long. We now have a platform from which to educate the masses in a way that is neutral and morally acceptable.

This debate is also the root of the separation of Church and State issues that have hindered education for decades. It is time to put these issues to rest and bring about a united effort to progress rather to remain in this stagnant debate and pass it on to yet another generation.

For Our Educators

Curricular materials are currently under development including primary and secondary texts, teacher's guides, student study guides, and test banks. We welcome and encourage your suggestions and other input to facilitate a truly universal curriculum.

Please direct your inquiries to:
Education@cometogetherbooks.com

Also under development are curricular materials specific to the specific needs of primary and secondary Christian educational institutions. Suggestions, comments, and your other inputs are encouraged and welcomed.

Please direct your inquiries to:
Christianeducation@cometogetherbooks.com

For Fellow Researchers & Scientists

It is the author's mission, through the creation, production, and distribution of easy-to-comprehend explanations for various phenomena, to create a true and universal understanding for the widest audience possible. It is only after such a task is accomplished that science can hope for a universal acceptance of ideas throughout society that can reduce the occurrence of circular and endless debate and allow a unified effort toward progress to occur.

If you are working on a project and would like to explore the opportunities that a "Come Together" type project might provide for a universal understanding of your ideas and theories contact the author for an analysis.

Please direct your inquiries to:
Projects@cometogetherbooks.com

Sincere Thanks to the Following

I have watched many colleagues in academia struggle with what they are teaching others and what they believe through their faith. I have also witnessed students strong in their faith reluctantly learn various theories for the purpose of passing an exam only to discard them later and feel guilty for being unfaithful to their religion and beliefs. It is exactly for these people, whom I once could count myself among, that I was inspired to create this work. That is also the purpose for my decision to start with the most controversial of subjects: a joining of Creation and Evolution.

I would like to express my sincere thanks to God and the following people who helped to make this work Universally Understandable and Universally Acceptable:

First is God, who gave me many of the words within and guided my message. He also permitted me enough relief from my physical disability to the extent necessary to continue this work.

My family, my lovely wife Laura and my Mini-Me Joe, who gave me the peace and quiet just when I needed it most as well as invaluable input.

My brother, David Hamilton, whose input was invaluable and led to the inclusion of more extensive illustrative and supportive examples through his favorite words to me: Prove it!

My best friend in faith, Quin Fortune, whose great faith in God I admire. Her non-scientific background and faith-based approach gave me great insight into what would be required for a universal understanding of my ideas on both sides of the issues.

I also have to thank all the great instructors in my life that truly made an impact upon my thinking and my life: Pastors James Kasler and Craig Groeschel, Dr. Steve Shore, Dr. Riaz Ahmad, Dr. Peggy Guthrie, Dr. Donna Zanowiak, Dr. Malinda Green, Dr. Myron Pope, Dr. William Caire, and the late Dr. Peter Wright.

A special thanks to Dr. Patrick Lo, D.O. who through his tireless dedication to science and medicine inspired me to pursue a similar course.

Finally my parents, Jerry and Charlotte, I know you are both cheering me on to greater things from the heights of Heaven. I hope you both are proud and pleased.

Terms

This section contains specific definitions of certain terms used within this book. The definitions presented are specific to the context of the terms within this book. These definitions are only provided to aid in a more thorough understanding of the concepts presented within this book. For more concise definitions the reader can refer to any dictionary desired.

Biblical Support or Test – Finding relevant passages that agree a relationship exists.

Creation – Pertains to the Biblical account of Creation events contained in Genesis the first book of the Bible.

Evolution – Any type of change usually occurring over time.

Experiments – Manipulation of an object or organism or its environment to induce a response.

Faith – The act of believing in the existence of a relationship, object, organism, or idea without any tangible or reproducible evidence that it exists.

Hypothesis – An idea or preconceived notion as to the results of observations and/or experiments.
(Plural: Hypotheses)

Indigenous – Anything that can trace substantial ancestry to the same general location.

Intentional & Intelligent Design – Pertains to the idea that an intelligent form – God – designed the events and contents of the Creation with specific and intended purposes for each.

King James Version – (KJV) Commissioned by King James this is the English translation of the Bible presented in 1611 A.D.

King James Version (Authorized) – (AKJV) A version of the King James Bible that includes certain words, identified in italics, to aid in comprehension of the Early English that the King James Version was written in. Also included are translations of harder to understand words or groups of words.

Law – Refers to theories that have withstood many attempts to refute them. Always specifically defined. Usually refers to behavior of objects.

Nature of Science – Referring to the ever-continuing search to explain various phenomena in understandable and reproducible terms.

Observations – Any act or behavior presented by an object or organism that is studied and can be recorded.

Parable – A translation of the Hebrew word 'mashal' which refers to illustrative speech or writing. Not to be taken in a literal sense it serves to convey a message or point of view.

Refute – Results of observations and/or experiments that deny a relationship exists.

Reproducibility – This allows that the results of observations or experiments will be repeated without regard to the time, place, or person performing the same observations or experiments.

Results – The data or information collected from observations and experiments.

Scientific Method – Detailing the conditions under which observations and experiments are performed in order to ensure reproducibility.

Scientific Support or Test – Supporting results of observations or experiments that follow the rules of reproducibility.

Support – Results or information that agrees that a relationship exists.

Theory – A hypothesis (see Hypothesis) that has withstood repeated attempts to support or refute it.
(Plural: Theories)

Universal Acceptance – The general acceptance of a theory by the majority of people based upon their understanding of it. See also Universal Understanding.

Universal Understanding – Explaining new ideas and theories in terms that the vast majority of people are capable of comprehending.

Coming Soon

Come Together Projects

Come Together: Cloning and Stem Cells

Come Together: Science and the Holy Bible
Past and Present

Author's Other Projects

The History of Science

The History of Biology

Printed in the United States
35688LVS00004B/103